Douglas M. Bryce

TROUBLESHOOTING
A Guide for Injection Molders

TECH TRAX LLC
IPLAS

TROUBLESHOOTING

A Guide for Injection Molders

ISBN No. 0-9705196-0-5

Library of Congress Catalog No. 2001088849

International Plastic Laboratories & Services (IPLAS)

Printed in the United States of America

ACKNOWLEDGEMENTS

I would like to take a moment to thank and acknowledge all those wonderful people who helped make this publication a success:

- Dave Reid and Jim Warren from TechTrax for their continued enthusiasm and help towards catalyzing the project and weaving all the strings together. In particular, I wish to acknowledge Dave for his editorial work and coordination of the book's production.

- Tish Holbrook of Holbrook Graphic Design for the layout and compilation activities.

- Shelley Readel of the Plastics Technology Deployment Center at Penn State Erie for her pre-production assistance.

And last, but not least, my wife Linda for her magnificent support and for putting up with a grouchy old plastics engineer while this project came to fruition.

Thank you all (or as they say here in Texas, y'all) for your efforts. Words alone can't express my appreciation for a job so well done. You are all considered experts in my eyes and friends in my heart.

Douglas M. Bryce
IPLAS
605 Ridgewood Rd. West
Georgetown, TX 78628
www.iplas.com

TABLE OF CONTENTS

INTRODUCTION

This book, and a series of companion products, has been created to provide a common sense approach for troubleshooting the thermoplastic injection molding process. The standard method of consulting 8-1/2 by 11" troubleshooting guides (as prepared by the material suppliers for specific materials) does not help enough to determine the cause of a defect in any specific thermoplastic material. This book and its companion products were created in a manner that will provide a basic understanding of what causes defects and the most effective ways of finding solutions to them. The purpose of this approach is to give you ideas to consider when you need to troubleshoot. I want you to understand the relationships between machine, mold, material, and operator. Then, you will be able to work your way through a problem situation using a systematic method, usually without the need to consult a guide of any sort. In fact, I want you to learn the troubleshooting process to the degree that you will never have to read this book again. It is written with the novice as well as the expert in mind. Actually, anyone involved with injection molding will find this book an asset, including supervisors, sales personnel, managers, technicians, operators, mold makers, material suppliers, engineers, quality personnel, and anyone else interested in the subject of injection molding thermoplastic materials.

This guide is written with a series of discussions on various topics that are followed by a separate chapter assigned to each of the most common defects found in the injection molding industry. The causes and remedies for each defect are listed in order of probability and are classified by Machine, Mold, Material, and Operator. You may notice that the same remedy will be used to solve a variety of defects. This is to be expected after you understand the interactions of the processing parameters and their many variables.

I hope you enjoy your experiences using this guide, and I suggest you investigate the companion products. There is an interactive CD, a small pocketbook version, a wall chart for quick reference, and other items you can research on our web sites located at www.texplas.com, www.iplas.com, and on our companion web site at www.techtrax.net.

Douglas M. Bryce

Chapter 1
GENERAL TROUBLESHOOTING

John Wesley Hyatt, along with his brother Isaiah invented plastic injection molding in 1868 in response to a need for replacing carved ivory for billiard balls. Since that time, the industry has evolved from an art to a science, especially with the advent of computers, the sophistication of molding equipment, and the emphasis on proper training at all levels.

Yet, there are many engineers, technicians, operators, supervisors, managers, sales personnel, etc., who are not able to troubleshoot a molding problem properly when defects occur. This book and its companion products were developed for those individuals, as well as any one else seeking to further their knowledge concerning defect solutions. Our approach is a little different. We don't expect you to study and commit to memory a bunch of rules and guidelines. There are plenty of those available through troubleshooting guides from material suppliers and machine manufacturers. Rather, we will take the approach of understanding what makes defects in the first place, then determine what solution will fix the problem and maybe even keep it from occurring again.

Troubleshooting Defined

For our purposes, *troubleshooting* can be defined as an activity that takes place to determine the cause of, and solution for, defects found in a molded part. This activity usually takes place while parts are being molded and occurs when the normal production of acceptable parts is interrupted by the unexpected production of one or more defective, unacceptable parts.

The person performing the troubleshooting activity may be an operator, a process technician, or a plastics engineer. For our purposes, we will call that person the "troubleshooter," although that actual title seldom appears as a job description.

In some cases, troubleshooting occurs when analyzing parts that were previously molded, such as when parts are returned from the field because they did not properly perform their intended function. Usually this situation is analyzed using failure analysis activities but troubleshooting may be also called upon.

What Causes Defects?

A study that took place over a 30-year span (by Texas Plastic Technologies) analyzed the root causes of most common injection molding defects. The defects studied were process related and did not include those resulting from poor product design. Everyone agreed to the product design with no further changes, and the molds were producing acceptable parts in a manufacturing environment. This study found that the defects could be traced to problems with one or more of the following four items: the molding machine; the mold; the plastic material; and the molding machine operator. The most interesting thing was what percentage each of these items contributed toward the cause of the defects, as shown in Figure 1-1.

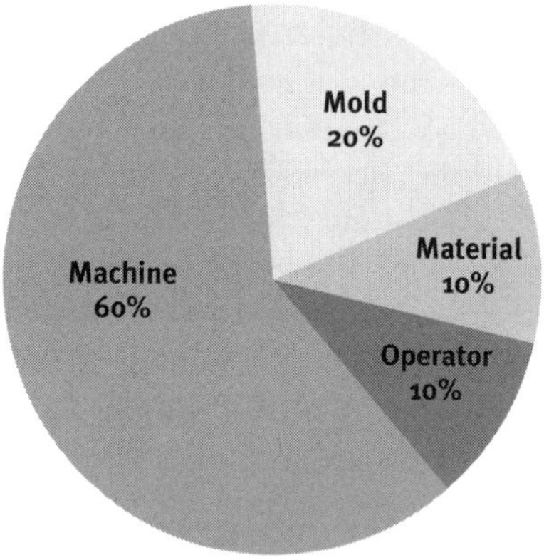

Figure 1-1 *Common cause of defects in molded parts.*

As Figure 1-1 shows, the most common cause of defects in molded parts is the molding machine itself, which is 60% of the time. This is followed by 20% of the defects caused by the mold, 10% caused by the material, and only 10% caused by the operator.

Contrary to what the chart shows (and the study found), most of us in the industry have held the belief that the most frequent cause of defects is usually the material, with the operator coming in a close second. However, as the chart shows, the actual most frequent cause of defects is the molding machine, followed by the mold. Thus, when troubleshooting, the first place to look for a solution to a defect problem is the machine because the answer will be there 6 out of 10 times.

Troubleshooting Tips

A troubleshooter must be able to approach a problem with an objective mind. What solved a problem one day may not solve the same problem another day. Because of the large number of parameters (more than 200) involved in molding, the variables of these parameters, and the way they all interact, many solutions may exist for a single problem. Conversely, many problems may be fixed using a single solution. So, the troubleshooter must think through the problem and make sure the most obvious solution is being utilized first. This is accomplished by applying objectivity, simple analysis, and common sense.

By visualizing what happens to the plastic as it travels from the hopper through the heating cylinder, and through the flow path to the cavity image, you can determine what may have changed that could be causing defects. A heater band could be burned out, an injection pressure valve spring may be weak, or cooling water lines may have become blocked. Any of these problems will cause specific things to happen. A thorough understanding of the molding process will help determine these problems and their solutions.

For now, it is important to understand that most defects can be corrected by one of a variety of changes. For instance, either drying the material, reducing nozzle temperature, or increasing the backpressure usually can correct splay. The trick to proper troubleshooting is to know when each solution will work and how to identify which problem is actually causing the splay. Understanding this concept will make the troubleshooting process less of a mystery and more of a science.

Successful troubleshooting usually requires making a change to an existing process. These changes will sometimes have an immediate effect, but in all cases, any changes will also have long-term effects. This is because the total molding process requires a certain amount of time to stabilize after any change is made. For instance, an increase in barrel temperature will alter the flow rate of a material after only a few minutes. However, that increase also has an effect on the injection speed after a few more minutes because the material is easier to move. A faster injection speed may initiate a tendency for flash to begin after a few more minutes.

There are three major considerations to follow when adjusting molding parameters:

1. Create a mental image of what should be happening during the process and look for obvious differences;

2. Make only one change at a time;

3. Allow the process to stabilize for a period of 10 to 20 cycles after any single change is made to the process.

If a particular change did not fix the problem, the change should be reversed (the setting put back where it was) before any other change is attempted. In addition, the next change should not be made until another 10 cycles have transpired. This demonstrates that troubleshooting can be a time consuming process because of the amount of time required to allow the machine to stabilize between changes. However, without that stabilizing time, so many changes can be made so quickly that no one could determine which change actually solved the original problem. Of course, the major concern is that the entire process may quickly go out of control because changing one parameter will affect another.

Our approach to troubleshooting is a little different from standard methods. We do not expect you to study and commit to memory a bunch of rules and guidelines. There are plenty of troubleshooting guides available from material suppliers and machine manufacturers. Rather, we will take the approach of understanding what makes defects in the first place, then determine what solution will fix the problem, and maybe even keep it from occurring again. You will not have to memorize anything, but we will expect you to learn where to find specific information such as recommended melt temperatures and mold temperatures. We will even include some of the more common information requirements in various locations throughout this book.

It is a common misconception that a troubleshooter must have years and years of experience, and had to absorb some amount of "black magic" over those years to solve processing problems. In fact, a good troubleshooter needs only a good understanding of the molding process, tools, and materials being used.

Too often a technician, engineer, or operator will be presented with a molding defect and will start turning dials, flipping switches, and adjusting timers without understanding what they are doing or knowing what results to expect. This is a common problem and is due to the fact that most troubleshooting is done as a result of doing something (anything) that worked in the past. Also, due to schedule requirements, a quick fix is desired, and the technician is pushed into a management-directed panic mode. The result is pandemonium as attempts to correct defects only seem to make matters worse and the entire molding process quickly goes out of control. While this is a standard scenario in most molding companies (but not highly publicized nor recognized), it does not have to be that way. The situation should be such that the troubleshooter can objectively analyze a molding defect and determine a probable solution before making any changes. The solution should be attempted, followed by another decision. Each solution should be determined independently and rationally. There should be no guesswork, and, when necessary, assistance from outside sources should be solicited and welcomed.

Developing Product Dimensions

One very common fault that occurs throughout our industry is that of trying to force the process to provide a molded part to specific dimensions when trying out a brand new mold. Although we can use the process parameters to make a part bigger or smaller, this should not be done on a new mold. This is because it minimizes the process parameter window for making changes when we do have the need. For example, when a new operator is placed at the machine, or the immediate environment changes, or a new batch of material is being processed the process parameters may need adjusting. It is better to create a process that produces the best visual part we can make, in the fastest cycle that we can safely attain. Then, once the process has stabilized (which may take up to eight hours), and the part has properly cooled (which may take another eight hours), we should check the part dimensionally. If any dimensions need adjusting, the mold should be pulled and the dimension adjusted by the moldmaker who built the mold originally. This is called *developing the mold* and should be considered standard practice for new molds.

Of course, once the process has been optimized and the mold has been developed and accepted for production, any future problems resulting in dimensional changes can be corrected with minor process adjustments. That is where troubleshooting comes in. An example would be excessive shrinkage from a change in melt index of the plastic material. A process adjustment can be made to accommodate this problem. That is why we need the large process window in the first place.

A more realistic proposal to troubleshooting is to use a two-pronged approach that consists of using the material suppliers' guides coupled with good old common sense. Remember: keep the troubleshooting process simple. In most cases, a single solution will correct a defective situation. The key is to come up with that single solution without wading through a myriad of possibilities.

So, a troubleshooter must be objective in analysis, selective in solution, and most of all, patient in activity.

Chapter 2

UNDERSTANDING THE MOLDING PROCESS

The injection molding process can be compared to filling a jelly donut. A machine prepares a plastic material (the jelly) and injects it into a closed mold that contains a vacant area (like the donut). The molding machine has two basic functions. First, it must prepare the plastic material using heat to melt the plastic, and inject that molten plastic into a closed mold using hydraulic pressure pushing against a plunger device. Second, it must generate enough clamping force to hold the mold closed against the amount of injection pressure being used to fill the mold. When these two activities have taken place, the mold is held closed until the molten plastic cools enough to form a solid skin on the molded part. Then, the part can be ejected and the next cycle can begin, as shown in Figure 2-1.

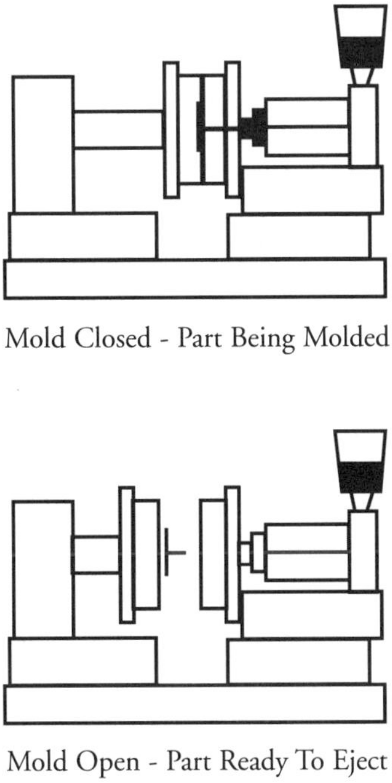

Figure 2-1 *Injection molding concept.*

PARAMETERS

There are over 200 different parameters that must be established and controlled to achieve proper injection molding of a plastic part. These parameters fall within four major areas: pressure, temperature, time, and distance, as shown in Figure 2-2.

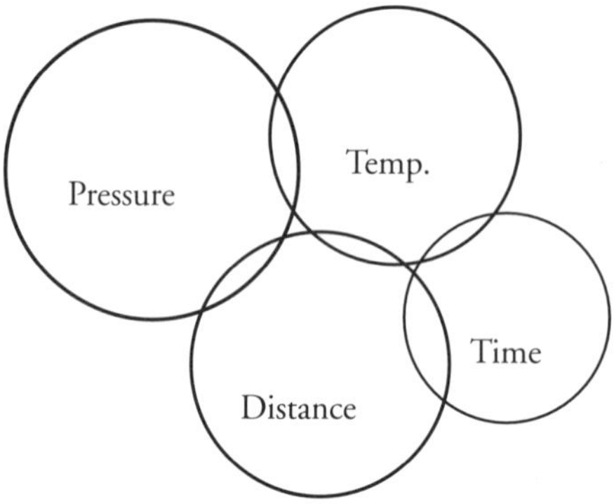

Figure 2-2 *Four parameter areas.*

To the troubleshooter, all four areas are important. But, the pressure and temperature areas are the ones most commonly considered during the troubleshooting process. Based on the requirements of any particular plastic material, the pressure must be sufficient to inject the plastic material and to hold the mold closed. In addition, the temperature of the injected plastic and mold must be correctly maintained.

In the following section, we will discuss the importance of various facets of the pressure and heat parameter areas. Distance and time will not be discussed here as they are beyond the intended scope of this chapter.

Pressure

Pressure is found primarily in the injection area, but there is also pressure found in the clamp unit of the molding machine. We will address all of these pressure requirements here.

BACKPRESSURE

The first pressure to consider is *backpressure.* This is pressure that is created during the return action of the screw after injecting material. The screw turns (augers) to bring fresh material into the heating cylinder. This material is placed in front of the screw and nudges the screw backwards. A buildup of pressure is created at the front end of the screw. This pressure is used for better mixing of the plastic (especially if colors are added at the press), removing small amounts of trapped air, and controlling the weight of the shot by maintaining an accurate density of a given volume of melt. The back pressure setting should start at 50 psi and be increased in 10 psi increments as needed, with a maximum setting of 300 psi. The maximum setting is needed because anything over that will cause too much shearing of the plastic and result in thermally degraded plastic.

INJECTION PRESSURE

The next type of pressure to consider is *injection pressure.* This is the primary pressure for injecting 95% of the molten plastic into the closed mold. Normally, the highest pressure and fastest fill rate are the best condition. However, high pressure will increase molded-in stress. And, that stress will be released at some time. There is no question as to its being released, only as to *when* it will be released. The greater the pressure, the greater the stress, and the greater the reaction when it is released. So, you should determine the minimum amount of pressure necessary to fill the mold, and then use all of it. And, remember, the hotter the plastic, the more fluid it becomes and the lower the pressure can be to fill the mold.

HOLD PRESSURE

Once the majority of the plastic (95%) has been injected using standard injection pressure, the machine should drop into hold pressure. This pressure is about half of the injection pressure and is used to finish filling the mold by packing the molecules together in an orderly fashion. Hold pressure is required until the gate freezes off, normally in 3 to 4 seconds. Once that happens, hold pressure has no more effect on the molecules on the other side of the gate. If hold pressure is released before the gate freezes, the material in the cavity is still molten and will be sucked back out of the cavity. At the very least, there will be insufficient pressure to pack the molecules together and uneven shrinkage and cooling will take place. If valve gating of a hot runner system is used, holding pressure can be released earlier than with standard surface gating.

CLAMP PRESSURE

At the other end of the machine, we have clamp pressure. The only reason to have clamp pressure is to keep the mold closed against injection pressure. Therefore, the amount of clamp pressure required is based on the material being molded. The easier flow materials require less injection pressure, thus they require less clamp pressure. Conversely, the stiffer flow materials will require more injection pressure, thus more clamp pressure.

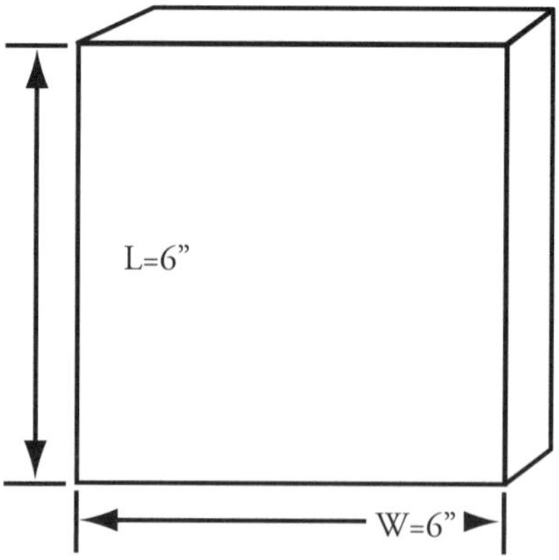

To determine how much clamp force is needed for a specific product, find the projected area of the part being molded and multiply it by two to six tons for each square inch of projected area.

Projected Area = L x W
(6" x 6" = 36 sq. in.)

***Figure 2-3** Determining clamp force.*

For example, if we were molding polycarbonate for the part shown in Figure 2-3, we would need five tons per square inch of area (as published by the polycarbonate supplier). So, the total force required would be 180 tons (36 sq. in. x 5 tons). But, if we were molding nylon 6/6, we would only need two tons per sq. in. (as published by the material supplier) so the total force would be only 72 tons (36 sq. in. x 2 tons). Therefore, we could run the mold in a smaller press, or use only a portion of the tonnage available in the press the mold is in. Reducing the tonnage requirement also reduces the cost to mold the part because we use fewer resources.

There is one factor, however, that must be discussed regarding clamp tonnage calculations. When the material suppliers state a specific clamp tonnage value (such as five tons per square inch for polycarbonate), the supplier assumes there is a proper shutoff land surrounding the cavity image. A typical shutoff land is shown in Figure 2-4. This shutoff land is simply a wall of steel surrounding the part and is approximately 0.002/0.003 inches in height and at least 1/4" wide.

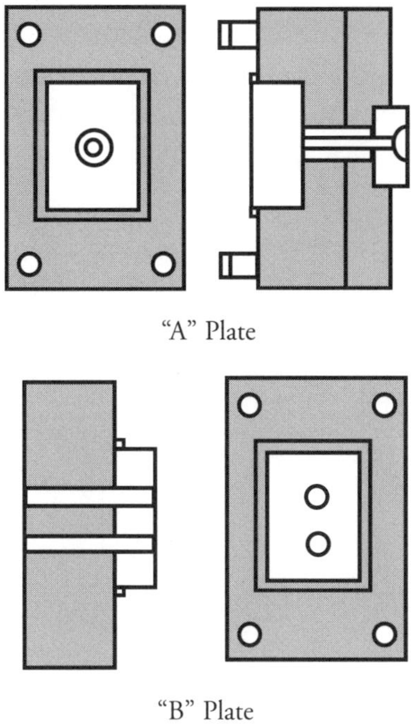

Figure 2-4 *Typical shutoff land.*

The purpose of this shutoff land is to concentrate all the clamp force to the area surrounding the cavity. That allows us to use less total force than if the clamp force is dispersed over the entire face of the mold base. Without the shutoff land, the amount of clamp force will be 3 to 4 times as much as with the shutoff land. The calculations shown earlier assume there is a shutoff land. If you use these calculations and the mold flashes easily around the part, chances are there is no shutoff land being utilized.

Heat

The next parameter area we will look at is heat. Heat is used to soften the plastic to the point of being able to inject it, but heat is also found in the mold and in the heat exchanger of the machine. We willl investigate controlling all of these areas.

INJECTION UNIT

There are four zones of heat that must be controlled in the injection unit. They are the rear, center, front, and nozzle zones. Each is controlled independently of the others. See Figure 2-5 for reference.

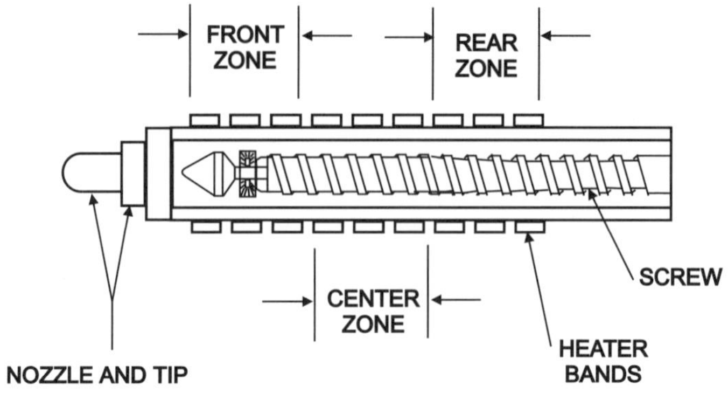

Figure 2-5 *Injection unit heat zones.*

The injection unit is designed to drag material through the four zones and to heat it gradually as it travels through the heating cylinder. The heat should be lower in the rear than in the front, and the nozzle should be the same as, or 10 degrees (F) hotter than, the front zone. Modern temperature controls are able to hold a temperature setting to within 1 degree (F).

It is important to understand, however, that the heat is created by heater bands strapped around the outside of the injection cylinder as well as by the turning action of the screw inside the heating cylinder. The heat is provided at a rate of approximately 50% for each of these devices.

The temperature control units do not measure the actual temperature of the plastic being heated, but rather the temperature of the steel of the injection unit's heating cylinder. This is important because the settings for the heating cylinder temperatures must be made higher than the actual melting temperature of the plastic. This is because the plastic is moving through the heating cylinder at a fairly rapid pace and must absorb enough heat during its travel to get to the desired temperature. In most cases, the heating cylinder will be set at 50 to 100 degrees (F) higher than the melt temperature of the plastic.

The plastic must be at the right molding temperature as it leaves the nozzle of the molding machine. We can determine that temperature by sliding the injection unit back from the mold and injecting material into the air. This is called an *air shot.* When it falls upon a special plate of the machine that is designed to accept this material (called a *purge plate*) we can use a pyrometer to measure the molten plastic temperature. Figure 2-6 depicts this action.

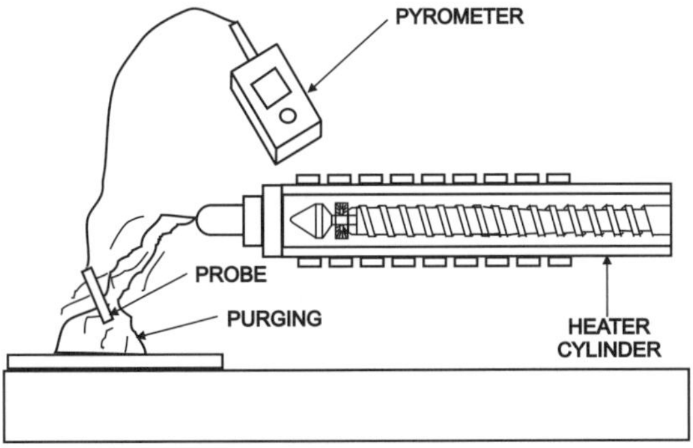

Figure 2-6 *Measuring material temperature.*

The temperature at this point should be the temperature requirement published by the material supplier. You can find this value for some common materials in Chapter 6 of this book.

MOLD TEMPERATURE

The most common method used for cooling the plastic once it is injected into the mold is a set of water lines. These lines are connected to a source of temperature-controlled water that circulates through the mold and pulls out heat that is building up in the mold over time. One common mistake of most troubleshooters is believing that the water leaving a mold should be hotter than the water entering the mold. The belief is that the water is used to pull heat from the plastic and therefore must be hotter when the job is complete. Actually, the water is being used to *maintain* the temperature of the mold and should be the same temperature leaving as entering. That is the definition of maintaining the temperature.

The waterlines are sized and located such that the water pulls heat out of the mold as fast as it is being generated. That is true maintenance of temperature. If the water leaving the mold is hotter than that entering the mold it means there is still a lot of heat left in the mold and the waterline design is not adequate to pull heat out as fast as it is being generated.

The material supplier defines the proper mold temperature and this is published information (see Chapter 7 for mold temperature requirements for some common materials). It must be cool enough to solidify the plastic quickly, but warm enough to keep it from becoming solid too fast. If the plastic solidifies too fast, the molecules do not have a chance to "bond" properly and the part will be weak or brittle. This is especially true of crystalline (or semi-crystalline) materials. Each plastic family has a specific mold temperature range within which it should be processed for highest quality parts at lowest possible cost.

HEAT EXCHANGER

The heat exchanger is a sophisticated radiator device that controls the temperature of the hydraulic oil used in the molding machine. This oil usually must be maintained at a temperature between 100 and 125 degrees (F) for proper use. If it is too cool, the machine actions are sluggish and inconsistent, and if it is too hot, the additives in the oil will fall out of solution and clog hydraulic mechanisms causing them to be inoperative or slow to respond. The oil is passed over a series of copper tubes that have water running through them. As with the mold, this circulating water pattern is designed to pull heat from the oil as fast as it is being generated. If the water leaving the heat exchanger is hotter than the water entering it, this indicates a blockage (such as calcium deposits) in the heat exchanger that is interfering with proper heat transfer. The heat exchanger must then be removed and the copper tubing must be flushed with acid or drilled out with special cleaning equipment.

CONTROL AND CONSISTENCY

The two most important things necessary for the molding of the highest quality products at the lowest possible cost are control and consistency.

Control should be applied to every possible parameter that can be determined. These include those of the four areas mentioned earlier, but also such items as the environment in which the parts are molded, the placement and operation of portable cooling fans, standardization of mold components, operation of secondary equipment, and any other action that is adjustable or variable. There should be no random or wandering motions or actions.

Consistency should be applied wherever possible, but primarily in machine actions. One area of major importance is that of any operator-controlled activity such as the opening and closing of a safety gate (often called the operator's gate). Most machine actions are controlled through electronic devices or computer controls and are extremely consistent. But any operator-controlled action tends to be inconsistent due to the nature of human endeavor. The human being works at varying speeds and distances throughout the day, ranging from quick and short at the start of a shift and after each break or rest period, but slower and longer in between those times. The inconsistent action causes inconsistent cycles and molded parts that vary in quality, not to mention cost. This is the major reason many molders are utilizing robots or other automated methods of molding. They wish to remove the human tendency towards inconsistency thus improving quality and reducing cost of the molded parts.

Human consistency can be achieved, however, through proper training and knowledge transfer. The operators can be instructed in the value of consistent operation and strive for consistency in their personal actions. A conscientious operator (and most are) will find ways of achieving the required consistency by humming tunes, tapping feet, singing songs, reciting poetry, or simply listening for a specific sound produced by the molding machine or mold. When they find a method that works they can be as consistent as the automated molding machine for an entire shift. Once control of as many parameters as possible is established and maximum consistency is achieved, defective parts will be minimized and production costs will be reduced as the quality of the molded parts increases. All of this reduces the necessity of troubleshooting activities as fewer defects are being produced.

Chapter 3
VENTING THE MOLD

Venting is a process that is used to remove trapped air from the closed mold and volatile gases from the processed molten plastic. Without venting, the trapped air will compress as plastic tries to force it out of the mold and the air will ignite, burning the surrounding plastic and causing charred areas on the molded part. Trapped air also keeps the plastic from filling in those areas of the cavity where the air is trapped so a non-filled (or short) part is molded. Volatile gases will be absorbed by the plastic and will cause voids, blisters, bubbles, and a variety of other defects.

The concept of venting is simple: provide many pathways to allow trapped air and volatile gases to escape from the mold quickly and cleanly. The pathways should lead directly from the edge of the cavity image of the mold, or through ejector and/or core pin clearance holes, to the outside atmosphere surrounding the mold. These pathways need to be deep enough to let air and gases out easily, but not deep enough to allow the molten plastic to escape through them.

Venting should be performed on every mold and the vents should be inspected periodically to make sure they are not blocked due to scale buildup. The scale can build up especially if the vents are not properly designed. The correct design is discussed in the next section.

VENT DIMENSIONS FOR COMMON MATERIALS

Depending on the type of plastic being molded, and whether it is stiff or easy flow, the vent depths should range from 0.0005" to 0.0020" in depth. Table 3-1 gives average depths for some common materials.

Table 3-1
Vent depth for common materials.

MATERIAL	CAVITY	RUNNER
ABS	0.002	0.004
ACETAL	0.0007	0.0015
ACRYLIC	0.002	0.004
CELLULOSE ACETATE	0.001	0.002
CELL. ACETATE BUTYRATE	0.001	0.002
IONOMER	0.0007	0.0015
NYLON 6/6	0.0005	0.001
POLYCARBONATE	0.002	0.004
POLYETHYLENE	0.001	0.002
POLYPROPYLENE	0.001	0.002
POLYPHENYLENE OXIDE	0.002	0.004
POLYPHENYLENE SULFIDE	0.0005	0.001
POLYSULFONE	0.001	0.002
POLYSTYRENE	0.001	0.002
RIGID PVC	0.002	0.004

The vent acts like a window in a wall. When the mold is closed the vent appears as a small tunnel going from the cavity to the edge of the mold. The dimensions for the vent are shown in Figure 3-1.

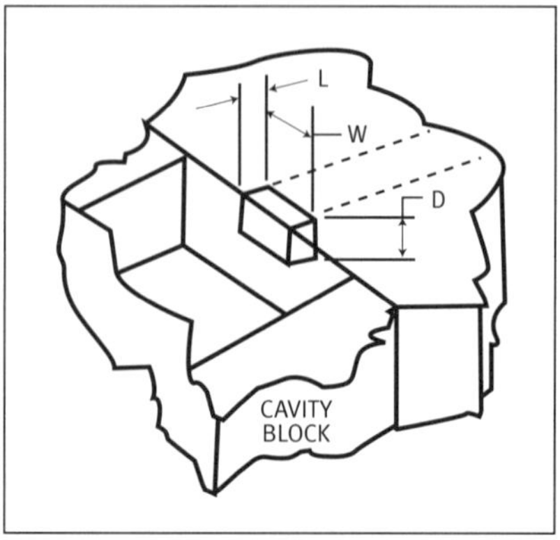

Figure 3-1 *Typical vent dimensions.*

There are three major dimensions for the vent. First is "D", the vent *depth*, already determined by the chart shown in Figure 3-1. Then, comes "W", the vent *width*. This can be anywhere from 1/8 inch wide or more, and a common width is 1/4 inch. There is no maximum width to a vent, but to be practical it should be somewhere between 1/8 and 1/2 inch in width. The final dimension is "L," which stands for *land*. This dimension should be a minimum of 0.030 inch and a maximum of 1/8 inch. If it is too short, the remaining steel is too weak and will not hold up. If it is too long, the tunnel shape of the vent is too great and the hot gases and trapped air will condense in the tunnel as they travel through the opening, and block the venting action. This will not occur in the first 1/8 inch of travel. Therefore, we make the "L" dimension a maximum of 1/8 inch and then open up the depth and/or width of the vent to allow the condensed gases to deposit beyond the vent itself.

HOW MANY VENTS ARE NEEDED?

The simple answer is that you cannot have too many vents. The more detailed answer is to have at least 30% of the perimeter of the cavity image in vents, equally spaced around that perimeter, as shown in Figure 3-2.

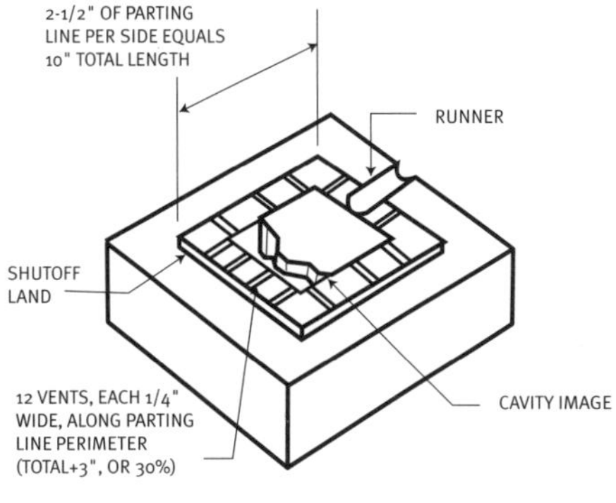

Figure 3-2 *How to determine the number of vents to put in a mold.*

Note that the part is small. It measures 2-1/2" per side. By measuring the perimeter of the part, we add up the sides to get a total of 10 inches. If 30% of that should be in vents, we need a total of three inches worth of vents. If we use 1/4 inch wide vents, we end up with a total of 12 vents for this part. They can be equally spaced as shown, or begin by placing a vent in each of the four corners, one directly opposite the gate, and the others equally spaced in the remaining areas.

Another, more conservative approach to venting states that there should be a vent at least every inch along the perimeter of the part cavity. In that case, there would be at least 10 vents for the part in Figure 3-2.

Do not forget to vent the runner. Any air or gas that is trapped in the runner gets pushed into the cavity and must be removed from there. One of the big advantages of venting is that, if properly done, it allows gas and air to get out quickly and helps the plastic to inject faster and at lower injection pressures thereby reducing stress and speeding up cycle times.

Chapter 4
UNDERSTANDING THE MATERIAL

The material that is used for injection molding (at least in the context of this book) is called thermoplastic. There is another type called thermoset, but that is another type entirely and the subject of a book by itself. We are talking about thermoplastic materials for our purposes.

Amorphous Versus Crystalline

Most of the materials you are working with can be classified in one of two orders, amorphous or crystalline in nature. There are major processing differences between these two groups and the properties of one are just about opposite of the other. First, we must understand the action of plastic molecules. Everything we know of is made up of billions of molecules. *Molecules* can be considered to look like miniature ping-pong balls. They are extremely small and can not be seen. They like to stay together and form a bond between themselves, giving a certain amount of strength to the item they are making. But, they do not like to be hot. So they try to move away from any heat that is applied to them. In fact, that is what you see when you look at boiling water. The steam is actually millions of molecules of water jumping out of the pot to escape the heat. Now, let's see how that applies to processing plastic materials.

Amorphous materials are those in which the molecular structure is random and becomes mobile over a wide temperature range. This simply means that the material begins to soften as soon as heat is applied. It gets softer and softer as more and more heat is applied until it becomes a liquid. It does not truly "melt" but we say it does anyway.

Crystalline materials are those in which the molecular structure is very ordered and becomes mobile only after being heated above its melting point. This means that the material does not soften as heat is applied. More and more heat must be applied until the material reaches a specific temperature (called its *melting point*) at which point it all melts at once. That is the real definition of melting. It happens all at once, unlike the amorphous material action.

Each material has a specific temperature at which it should be processed, and this is based upon the molecular activity of that particular plastic. The manufacturer of the material supplies this information, and we have listed some of the more common material temperature requirements in Chapter 6. The recommended temperature should be closely followed when processing a specific material. If the temperature is too high, there is a risk of degrading it. If the temperature is too low, there will be a need to increase injection pressure to force it into the mold and stress will be increased. Once the proper temperature is found, it is usually practical to stay within 50 degrees (F) of it, as measured when the material leaves the machine nozzle. Of course, some materials are very heat sensitive (such as PVC) and this rule-of-thumb may not apply in all cases.

Understanding Shrinkage

One of the most difficult properties of a molded part to predict is shrinkage. *Shrinkage* is the difference between the dimensions of a molded part when it is first molded, and after being removed from the mold and left to cool to room temperature. Molecules are expanded when exposed to heat. This is similar to the effect of heating the air in a balloon. The balloon will expand. Then, as the air in the balloon cools down the balloon will contract. The situation is the same with plastic molecules. They expand when heated and contract when cooled. We call that amount contraction *shrinkage*, and each material has its own amount of shrinkage.

Shrinkage is usually classified in three orders: low, medium, and high. Amorphous materials have low shrinkage, semi-crystalline materials have medium shrinkage, and crystalline materials have high shrinkage. *Low shrinkage* has a range of from 0.000 to 0.005 inch per inch. That means that every inch of the part will shrink that amount. So shrinkage is rated as so much "inch per inch" and is usually written as "in/in." *Medium shrinkage* has a range of from 0.005 in/in to 0.010 in/in. And, *high shrinkage* has a range that is anything above 0.010 in/in.

But, to complicate matters, there are two types of shrinkage: isotropic (pronounced ice-o-tropic) and anisotropic (pronounced ann-ice-o-tropic). Amorphous materials tend to display isotropic shrinkage, and crystalline materials tend to display anisotropic shrinkage. *Isotropic shrinkage* is shrinkage that is equal in all directions. *Anisotropic shrinkage* is shrinkage that is greater in one direction than another, usually in the direction of flow.

But, again to make matters worse, if reinforcement (such as fiberglass, nickel, graphite, etc.) is added to the plastic, all materials will shrink more across the direction of flow instead.

You may be able to affect the amount of shrinkage that occurs in a part by altering process conditions. For example, an increase in injection pressure will force the molecules together tighter and result in less expansion. Therefore less shrinkage. But, this should only be done as a temporary measure as it will reduce the total processing window that is needed for proper processing and make other changes almost impossible. If a dimension is always out of tolerance, the mold should be reworked to bring the dimension back in to a reasonable control level. While shrinkage is extremely unpredictable, it can be consistent once the process is under control, and an understanding of it will help in making specific troubleshooting decisions.

THE IMPORTANCE OF DRYING

All materials must be dry before they are processed. The reason for this is that any moisture present in the material will turn to steam in the heating cylinder. This steam will be pushed into the mold and will interfere with the molecular bonding that determines the strength of the molded product. The result will be a physically weak and visually unattractive, molded part.

Some materials are referred to as being hygroscopic (hi-grow-scaw-pick), while others are called non-hygroscopic. *Hygroscopic materials* act like sponges and absorb moisture directly from the atmosphere. This moisture goes right to the center of the molecules. Some of the common hygroscopic materials are ABS, polycarbonate, and nylon. The *non-hygroscopic materials* do not absorb moisture, but are still affected by moisture in the form of condensation that occurs in humid conditions. The moisture will sit on the surface of the plastic pellets. These materials include polypropylene and polyethylene. Another concern regarding non-hygroscopic materials is that although the basic plastic is not hygroscopic some of the fillers used to make the plastic may be hygroscopic and they will absorb moisture from the atmosphere.

It is common knowledge that hygroscopic materials must be dried to remove the moisture. But, the non-hygroscopic materials are usually dried only when moisture is known to be present such as in very humid environments. However, it is better to plan on drying all materials all of the time. It never hurts to dry material. It can not be too dry. It is possible to use too much drying heat, though. If you automatically dry the material, you will never have to be concerned as to when you should dry and when you do not have to. That will eliminate the problem that occurs if you do not dry and then find out you should dry only after trying to process the plastic. Then, you will have to shut down the machine for a few hours while you dry the resin properly.

How can you tell if the material is *dry* enough to mold? The most common method is to use a dew point meter to test the circulating air being used to do the drying. The dew point measurement should be from -20 to -40 degrees (F). You can also test by using a moisture analyzer. This machine weighs a sample, heats it to drive off the moisture, and then weighs it again. There should be no more than 1/10 of 1% moisture by weight to mold the plastic.

A word of advice: get in the habit of drying all materials whether they are hygroscopic or not. That way, you will always be prepared. And drying the material will pre-condition it thermally. It will always run better than non-dried material. Finally, the dried material must be used within two hours of drying, as it will begin to be affected by moisture again after that time.

Chapter 5
THE OPERATOR

The Role of the Operator

Although automation has taken a big step into the injection molding industry, the majority of molding operations still rely on an operator to run the molding machine. In most cases, the operators duties include opening the safety gate at the end of the cycle, removing the molded plastic (including the runner), closing the safety gate to start the next cycle, and removing the molded parts from the runner system.

The operator is the most important factor in the molding process because the operator is closest to the operation on a daily basis and usually is aware of any problems that occur before anyone else. In some cases, the operator is allowed to make process changes to correct some problems, and in other cases, the operator is not allowed to make any changes at all. This is dependent on how much training the operator has in processing and troubleshooting. Normally, the operator simply watches for problems and informs a supervisor or technician when trouble begins, and then the troubleshooting process starts. It is a good idea to involve the operator at this point by asking questions as to what happened just prior to the problem. For instance, the operator may have noticed that there was contamination in the batch of material that was recently placed in the hopper. Or, the operator may have heard a banging noise that might indicate a hydraulic component malfunction. The operator is a good source of information because he/she knows the job and the expected actions inside and out. Anything out of the ordinary is very obvious to the operator.

Influence On Cycle Consistency

Most people are not aware that it may take up to eight hours for a machine and the molding process itself to get stabilized after it is first started. It takes that long for the heats to stabilize, the hydraulic oil to come to proper consistency, and for the machine and mold to expand and optimize. During that startup period, it may be necessary to tweak the controls continuously to obtain acceptable parts. But, eventually the machine, mold, and process will equalize and become stable. Any change in the process or cycle time causes the machine to become unstable and it will take anywhere from 10 minutes to a full eight hours for the machine to re-stabilize (depending on the nature of the changes).

Because the operator has control only over the opening and closing of the safety gate, this part of the total cycle tends to be inconsistent. The rest of the cycle is timed and controlled by the molding machine and is extremely accurate from cycle to cycle. But the operator, as a human being, becomes less and less consistent throughout the shift. He or she will begin the shift fresh and in an upbeat mood. As the shift goes on, the operator begins to slow down and gets sluggish. Therefore, the overall cycle time slows down too. Then, after breaks and lunchtime, the operator is fresh again and works at a faster pace. Now the total cycle is faster too. This scenario repeats itself over and over throughout the entire shift. The inconsistent cycling of the machine causes it to become unstable and the parts being molded will be inconsistent in quality. This is one reason for defects that seem to come and go without anyone being able to discover the cause or solution.

It is important to explain to the operator the critical role that consistency plays in the production of parts with consistent quality. The operator may be able to achieve consistency by finding a way of maintaining a steady rhythm from cycle to cycle. This is an easy thing for an operator to do and takes the form of singing, whistling tunes, reciting poetry, tapping feet, or simply listening for telltale noises during the cycle. But, some operators can never achieve the consistency required, and some cannot perform the same way from day to day. In addition, another operator must relieve every operator for such times as breaks and lunch. The relief operator cannot be expected to work at precisely the same pace of the operator being relieved. This causes inconsistency in the cycle and the molded products.

Automated Alternative

Because of the inconsistency problem, many companies are replacing human operators with automated systems and robot operators. These robots can be programmed to be extremely consistent from cycle to cycle, never have to be relieved for lunch or breaks, and do not take vacations or time from work for other reasons. Use of robots takes away the human involvement and eliminates inconsistent cycles.

Even without robots, many molds can be designed to run in an automatic mode with the parts being molded and ejected (still connected to the runner or degated within the mold as it opens) without the need to open and close the safety gate. This results, again, in very consistent cycles and parts.

While these practices are intended to remove the human inconsistency factor from the molding process, they do not necessarily remove the human altogether. It is still a good idea to have a person watching over the process to interrupt the machine if a problem occurs and to act as an inspector to keep an eye on parts being molded. The job function could be altered to replace machine operation with inspection, carton making, packaging, and machine maintenance among other possible duties.

Chapter 6
MELT TEMPERATURES

Heating temperatures for plastic can be based on manufacturers' specifications or by testing.

Manufacturers' Specifications

The manufacturer of any specific plastic molding compound supplies data with each shipment of that material and provides processing recommendations designed to allow the molder to produce high quality molded parts in the fastest possible cycle time. Table 6-1 shows the recommended melt temperatures (as determined by the suppliers) for some common materials. Please remember to check the melt temperature as it leaves the nozzle. This is accomplished by pulling the sled back and purging to the air. Then, immediately plunge a pointed probe from a fast-acting pyrometer into the purged material and record the temperature. An average of three readings is recommended.

Table 6-1 *Recommended melt temperatures.*

Material	Degrees (F)	Material	Degrees (F)
Acetal (CoPo)	400	PBT	500
Acetal (HoPo)	425	PCT	580
Acrylic	425	PEEK	720
Acrylic (Mod)	500	PET	540
ABS (MedImp)	400	Polycarbonate	550
ABS (HiImpFR)	420	Polyetherimide	700
CelAcetate	385	Polyethylene (LD)	325
CelButyrate	350	Polyethylene (HD)	400
CelPropionate	350	Polypropylene	350
EVA	350	Polystyrene (GP)	350
LCP	500	Polystyrene (MI)	380
Nylon (6)	500	Polystyrene (HI)	390
Nylon (6/6)	525	Polysulfone	700
Polyamide-imide	650	PPO	575
Polyarylate	700	PVC (Rig/Flex)	350/325
		TFE	600

One of the pieces of information available is the recommended melt temperature. This is the temperature the material should be as it leaves the molding machine (through the nozzle) and enters the mold (through the sprue bushing). This temperature is different for every material made, and should be held as close as possible by the molder to attain parts with expected physical, mechanical, thermal, and electrical properties. Materials are chosen for how their combination of properties reacts to a specific product design requirement. The temperature at which the material 9is molded determines the property reactions. That is why the melt temperature setting of the molding machine is so important.

Testing

The method used to determine the proper molding temperature of any plastic is based on the use of test equipment called the *differential scanning calorimeter* (DSC). In a simplified explanation of the DSC process, a small amount (20 milligrams) of plastic material is placed in a chamber of the test unit and the DSC begins applying heat to that sample of plastic. The DSC must determine how much heat is required to get all of the molecules of the sample moving. It displays a graph showing this movement. Typical graphs for both amorphous and crystalline materials are shown in Figure 6-1.

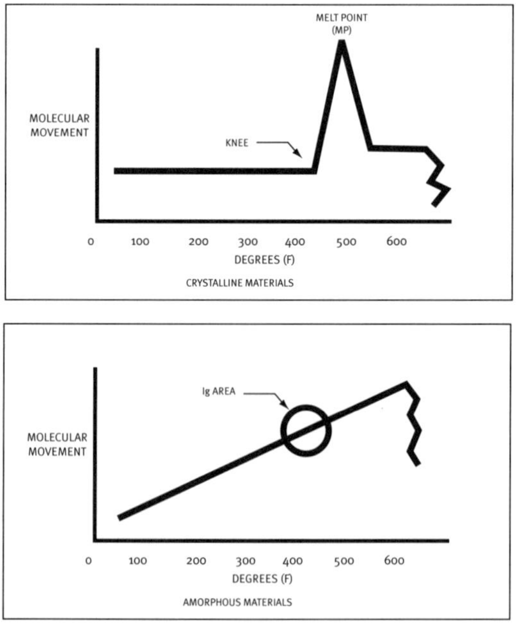

Figure 6-1 *Differential scanning calorimeter graphs.*

The amorphous graph shows that as soon as heat is applied the plastic molecules begin moving. The amorphous material goes through a series of phases during which it changes from a hard substance to a softer substance, finally to a liquid, and eventually degrades. At the point at which it became liquid the phase is called a *glass transition phase.* That is the temperature at which we should mold the material to obtain the highest quality level of property values.

For the crystalline material, the molecules do not begin to move until the temperature is close to what is called the *melting point* of the material. This is due to the molecular structure order of crystalline materials. When the maximum molecular movement is attained the material is at its melt point. That is the temperature at which we should mold the crystalline material.

These are the temperature values that the material supplier tests for when developing a material, and these are the values that get published in their data sheets.

Influence of Residence Time

Residence time can be defined as the total amount of time a material resides in the heating cylinder of the molding machine (with heat on) before it is injected into a mold. The material temperature settings on the barrel must be made determined by the residence time. The normal settings would be made based on 50% of the heating cylinder (barrel) being emptied every cycle. If we have a total shot size of three ounces (including the runner system) we should run the mold in a machine with a six-ounce barrel. This is an ideal situation because it means we are injecting one batch of material for a cycle while preparing the next batch for the next cycle. By doing this, the residence time is kept at an ideal value.

However, due to the variation in size of all the different parts we will mold, the amount of material being injected each cycle may not equal 50% of the barrel size for every production run. We use a rule-of-thumb that states we can use from 20% to 80% of the barrel capacity depending on the heat sensitivity of the plastic being molded. A material that is very heat sensitive (such as PVC) should be run in a machine using up to 80% of the barrel capacity every cycle or the plastic will degrade in the barrel due to too much residence time. But, a material that is thermally forgiving (such as polypropylene) can be run in a machine using only 20% of the barrel capacity every cycle because the material will not degrade easily due to excessive residence time.

Therefore, the melt temperature stated by the material supplier for a specific material is based on averages and 50% barrel capacity usage. It should be used as a guideline but you must be flexible in establishing the final melt temperature based on the conditions under which you are molding.

Chapter 7
RECOMMENDED MOLD TEMPERATURES

The temperature that you set for the mold can depend on manufacturer's specifications and the properties that you want for the final product.

Manufacturer Specifications

The material manufacturer or supplier is responsible for providing the recommended temperature of the mold into which we are injecting the molten plastic material. The theory followed is that we are heating a plastic up to a specific melting temperature, and then injecting it into a mold in which we wish the plastic to cool down to a solid again. However, every plastic has a specific rate at which it gives up heat. If we try to take heat away too quickly, it may not attain the maximum degree of physical, chemical, thermal, and electrical properties that are available. The manufacturer runs tests on each material developed to determine the best mold temperature needed to produce the highest quality parts.

When To Vary Mold Temperature

While the material supplier may recommend a specific mold temperature for a specific material, it is possible to run the mold at a higher or lower temperature as long as it is understood that property values will suffer more as you stray from the recommended temperature. For instance, if you are using a polypropylene material for molding disposable flower pots you may be able to run the mold at 60 to 80 degrees (F) which is 40 to 60 degrees lower than the recommended 120 degrees (F) because you do not need the high level of property values that are available. But, if you are molding electrical components from the same material you should not vary far from the recommended 120 degrees (F) because you need to attain the maximum electrical properties possible or you risk product failure in the field.

It is important to understand the final use of the product being molded to determine the range of mold temperature you can use. In any case, there should be no more than a 10-degree (F) difference between any two points of the steel area of the mold that actually forms the parts. If there is more than that difference, there will be a thermal shock condition that will induce a large amount of stress into the molded part. This can be found as warp, bowing, brittleness, and other similar defect conditions.

Table 7-1 shows the recommended mold temperatures (as determined by the suppliers) for some common materials. The mold temperature should be checked by using a surface probe on a fast-acting pyrometer and checking the area of the mold where the molten plastic will be formed. The other areas do not count. Please remember that water control units cannot effectively maintain mold temperatures above 190 degrees (F). You will need the assistance of cartridge heaters in the mold or the use of an oil control unit instead of the water control unit for those situations.

Table 7-1 *Recommended mold temperatures*

Material	Degrees (F)	Material	Degrees (F)
Acetal (CoPo)	200	PBT	180
Acetal (HoPo)	210	PCT	250
Acrylic	180	PEEK	380
Acrylic (Mod)	200	PET	210
ABS (MedImp)	180	Polycarbonate	200
ABS (HiImpFR)	185	Polyetherimide	225
CelAcetate	150	Polyethylene (LD)	80
CelButyrate	120	Polyethylene (HD)	110
CelPropionate	120	Polypropylene	120
EVA	120	Polystyrene (GP)	140
LCP	250	Polystyrene (MI)	160
Nylon (6)	200	Polystyrene (HI)	180
Nylon (6/6)	175	Polysulfone	250
Polyamide-imide	400	PPO	140
Polyarylate	275	PVC (Rig/Flex)	140/80
		TFE	180

Chapter 8
GLOSSARY OF COMMON TERMS

Following is a list of common terms found in our industry. Please note that it is not all-inclusive. Also, please note that some words have multiple meanings. Be sure you are using them in the right context.

Glossary of Terms

Additive. A substance added to a plastic compound to alter its characteristics. Examples are plasticizers, reinforcements, and flame-retardants.

Alloy. A combination of two or more plastics that form a new plastic. See *blend.*

Amorphous. A plastic material in which the molecular structure is random and becomes mobile over a wide temperature range. See *crystalline.*

Anneal. To heat a molded part up to a temperature just below its melting point and slowly cool it back down to room temperature. This relieves molded stresses. See *conditioning.*

Anisotropic shrinkage. Shrinkage that occurs more in one direction (usually the direction of flow; reinforced materials shrink more across the direction of flow) than another.

Automatic operation. The term used to define the mode in which a molding machine is operating when there is no need for an operator to start each cycle.

Barrel. A metallic cylinder containing the injection screw (or plunger). Also called *cylinder.*

Blend. A mixture of two or more plastics.

Boss. A projection of the plastic part, normally round, which is used to strengthen an area of a part; provide a source of fastening; or to provide an alignment mechanism during assembly.

Cartridge heaters. Pencil-shaped electrical heater devices sometimes placed in molds to raise the temperature level of the mold. Especially beneficial when molding high-temperature crystalline materials.

Cavity. A depression or female portion of the mold that creates the external plastic part surface.

Check ring. A ring shaped component that slides back and forth over the tip end of the screw. The check ring eliminates the flow of molten material backwards over the screw during the injection process.

Clamp force. The force, in tons, that the clamp unit of a molding machine exerts to keep the mold closed during the injection process.

Clamp unit. That section of the molding machine containing the clamping mechanism. This is used to close the mold and keep it closed against injection pressure created by the injection process. The clamp unit also contains the ejection mechanism.

Cold slug well. A depression (normally circular) in the ejection half of an injection mold, opposite the sprue, designed to receive the first front, or "cold" portion, of molten plastic during the injection process.

Compression ratio. A factor that determines the amount of shear that is imparted to plastic material as it travels through the barrel. It is determined by dividing the depth of the screw flight in the feed section by the depth of the screw flight in the metering section.

Conditioning. Exposing a molded part to a set of conditions (such as hot oil), which impart favorable characteristics to the product. See *anneal.*

Cooling channels. Drilled holes or channels machined into various plates or components of an injection mold providing a flow path for cooling medium (such as water) to control the temperature of the mold.

Core. A) an extended or male portion of the mold that creates the internal plastic part surface. B) a pin or protrusion designed to produce a hole or depression in the plastic part.

Counterbore. A recessed circular area. Commonly used to fit the head of an ejector pin (return pin, sucker pin, etc.) in the ejector plate.

Crystalline. A plastic material in which the molecular structure becomes mobile only after being heated above its melting point. See *amorphous.*

Cushion. A pad of material left in the barrel at the end of the injection stroke. It is excessive to the amount needed to fill the mold and acts as a focus point for holding pressure against the cooling melt.

Cycle. The total amount of time required for the completion of all operations needed to produce a molded part. Sometimes referred to as the *gate-to-gate* time, meaning the time from when an operator first closes the gate until the time the operator closes the gate again for starting the next cycle.

Cylinder. See *barrel.*

Decompression. A method of relieving pressure on the melt after preparing it for injection during the upcoming cycle. This minimizes the drooling that occurs when a shutoff nozzle is not utilized.

Defect. An imperfection in a molded part that results in the product not meeting original design specifications. These defects can be visual, physical, and/or hidden.

Draft. An angle (or taper) provided on the mold to facilitate ejection of the molded part.

Ejector half. The half of the mold that is mounted to the moving platen of the injection machine. Sometimes called the "live" half or the "moveable" half because it moves. This half of the mold usually contains the ejection system.

Ejector pin. A pin, normally circular, placed in either half of the mold (usually the ejector half), which pushes the finished molded product, or runner system, out of a mold. Also referred to as a *knockout* pin.

Feed throat. The area at the rear end of the injection unit that allows fresh plastic to fall from the hopper into the heating barrel.

Feed zone. The area of the screw that is at the rear of the injection unit and receives fresh material from the feed throat.

Filler. Specific material added to the basic plastic resin to obtain particular chemical, electrical, physical, or thermal properties.

Flash. A thin film of plastic that tends to form at parting line areas of a mold. May also be found in vent areas and around ejector pins. Flash is caused by too great a clearance between mating metal surfaces, which allows plastic material to enter.

Flight. The helical metal thread structure of the injection screw.

Gate. An opening found at the entrance of a cavity (end of the runner system) that allows material to enter.

Granulator. A machine designed to grind up rejected pre-molded plastic (products or runners). The material generated by this process is called *regrind.*

Guide pins. A pin (usually circular) that normally travels in a bushing to provide alignment of two unattached components, such as the two halves of an injection mold. Also called *leader pins.*

Heating cylinder. That section of the injection molding machine in which the plastic resin is heated to the proper molding temperature prior to injection into the mold.

Heating zone. An area of the heating barrel that is controlled by a temperature controller attached to a set of heater bands. There are four major zones: rear, center, front, and nozzle.

Heater bands. Bracelet-shaped electrical heaters that are placed around the outside circumference of the heating barrel.

Hopper. A funnel-shaped container mounted over the feed throat of a molding machine. It holds fresh material to be gravity fed into the feed zone of the heating barrel. Hoppers are normally designed to hold enough material to run the injection molding process for an average of two hours.

Hydraulic clamp. A term used to describe the use of a large hydraulic cylinder to open and close the clamp unit of a molding machine.

Hygroscopic. A term applied to those plastics (such as ABS and nylon) that absorb moisture from the atmosphere.

Injection capacity. A rating of the maximum amount of plastic material, in ounces, a machine is capable of injecting in a single stroke of the injection screw or plunger. It is based on the specific gravity of polystyrene as a standard.

Injection molding. The process of pushing a molten plastic material into a relatively cooled mold to produce a finished product.

Injection pressure. That pressure that performs the initial filling of the mold. It is supplied by the injection screw or plunger as it pushes material out of the heating barrel and into the mold.

Injection unit. The section of the molding machine that contains the injection components, including the hopper, heating cylinder, screw (or plunger), nozzle, and heater bands.

Isotropic shrinkage. Shrinkage that occurs equally in all directions. See *anisotropic shrinkage.*

Knockout pin. See *ejector pin.*

Land. A term used to describe the area in which the gate, or vent, resides. It can also be thought of as the "length" dimension in the "l, w, h" terminology used for describing the dimensions of the gate or vent. See also *shutoff land.*

L/D ratio. The result of a calculation that divides the entire length of flighted area on a screw by its nominal diameter.

Leader pin. See *guide pin.*

Manual operation. The term used to define the mode in which a molding machine is operating when there is a need for an operator to start and finish each phase of the total cycle.

Mechanical clamp. See *toggle clamp.*

Melt. A term given to describe the condition of molten plastic prior to injection into a mold. A proper melt has the consistency of warm honey.

Metering zone. The area of the screw at the front end that contains properly melted plastic, which is ready to inject.

Mold. The term given to the entire tool (cavity, core, ejectors, etc.) Needed to produce molded parts from molten plastic material.

Monomer. A molecular unit of an organic substance, usually in the form of a liquid or gas. See *polymer.*

Moving platen. The platen of a molding machine that travels (opens and closes). It is connected to the clamp unit and is the mounting location for the "b", or traveling half of the mold.

Non-return valve. A mechanism mounted in (or at) the nozzle of the injection machine, which operates to shut off injection flow at the end of the injection cycle. This eliminates material from the upcoming shot from drooling out of the nozzle when the mold opens to eject parts from the previous shot.

Nozzle. A device mounted at the end of the heating barrel that focuses plastic material to flow from the machine into the mold.

Pad. See *cushion.*

Parting line. A plane at which two halves of a mold meet. Also applies to any other plane where two moving sections come together and form a surface of a molded part.

Plastic. A complex organic compound (usually polymerized) that is capable of being shaped or formed.

Platens. The flat surfaces of a molding machine to which the two halves of the mold are mounted. One is stationary and the other travels. There is a third platen (stationary) at the clamp end of the machine that serves as an anchoring point for the clamp unit.

Plunger. The injecting member of a non-screw design molding machine. Plungers do not rotate (auger) to bring material forward in preparation for the next cycle. Nor do they blend the material as a screw does.

Polymer. A group of long chains of monomers, bonded together in a chemical reaction to form a solid. This term is often used interchangeably with *plastic*, but there can be a difference.

Purging. A process of injecting unwanted plastic material from the injection cylinder into the atmosphere for the purpose of changing materials, changing colors, or removing degraded material. Also, the name given to the mass of material that is purged.

Reciprocating screw. A helical flighted, metal shaft that rotates within the heating cylinder of a molding machine, shearing, blending, and advancing the plastic material. After rotating, the screw is pushed forward which injects the plastic into the mold. Also, simply referred to as "*the screw.*"

Regrind. Plastic material formed by granulating pre-molded material. Regrind is virgin material that has been exposed to at least one heating cycle.

Runner. Grooves or channels cut into either or both halves of the injection mold to provide a path for the molten plastic material, which is to be carried from the sprue to the gate(s) of the cavity.

Screw. See *reciprocating screw.*

Screw speed. The rotating speed of the screw as it augers new material towards the metering zone. It is expressed in rpm (revolutions per minute).

Secondary operation. Any activity performed after the molding process required to produce a finished product suitable for its designed purpose.

Semi-automatic operation. The term used to define the mode in which a molding machine is operating when there is a need for an operator to start each cycle.

Shot. A term given to the total amount of plastic material that is injected (or shot) into a mold in a single cycle.

Shot capacity. See *injection capacity.*

Shutoff land. A raised area of the mold surface surrounding the cavity image. This area is usually between 0.002 and 0.003 inch high, approximately 1/2 inch wide and is used to focus clamping pressure on the mold. The use of a shutoff land reduces the amount of tonnage required to keep a mold closed against injection pressure.

Slide. A section of the mold that is made to travel at an angle to the normal movement of the mold. It is used for providing undercuts, recesses, etc.

Sprue. The plastic material that connects the runner system to the nozzle of the heating cylinder of the molding machine. It is formed by the internal surface of a bushing that joins the mold to the machine's nozzle.

Sprue bushing. A hardened bushing that connects the mold to the molding machine nozzle and allows molten plastic to enter the runner system.

Stationary platen ("a"). The platen at the injection end of the molding machine that does not travel. It contains the "a" half of the mold and locates the mold to the nozzle of the injection unit. The moving platen travels between this platen and stationary platen "b."

Stationary platen ("b"). The platen at the clamp end of the molding machine that does not travel. The moving platen travels between this platen and stationary platen "a."

Stress. A resistance to deformation from an applied force. Molded plastic products tend to contain stresses molded in as a result of forces applied during the injection process. These stresses may result in fractures, cracks, and breakage if they are released during use of the product.

Suck back. See *decompression.*

Support pillar. A circular rod mold component used to support the ejector half of the mold. It is required because of the tremendous amount of pressure exerted against the "b" plate by the injection phase of the molding process.

Thermocouple. A device made of two dissimilar metals that are used to measure the temperature of a heated area such as a barrel or nozzle. It sends a signal to a controller, which then adjusts the temperature of that area.

Thermoplastic. A plastic material that, when heated, undergoes a physical change. It can be reheated, thus reformed, over and over again. See *thermoset.*

Thermoset. A plastic material which, when heated, undergoes a chemical change and "cures". It cannot be reformed, and reheating only degrades it. See *thermoplastic.*

Tie bars. Large diameter rods that connect stationary platen "a" to stationary platen "b." The moving platen contains bushings that are used for sliding over the tie bars, allowing the moving platen to travel between the two stationary platens.

Toggle clamp. A term used to describe the use of a mechanical "scissors action" system to open and close the clamp unit of a molding machine. It is operated by a relatively small hydraulic cylinder.

Transition zone. That area in the center of the screw (between the feed zone and metering zone). This section has a tapering flight depth condition which compresses the plastic material in preparation for injection.

Undercut. A recess or extension on the molded part, located in such a way as to prevent or impede ejection of the part by normal molding machine operation.

Vent. A shallow groove machined into the parting line surface of a mold to allow air and gases to escape from the cavity, or runner, as the molten plastic is filling the mold. Sometimes also located on ejector and core pins.

Vented barrel. A heating barrel designed with an automatic venting port that allows moisture and gases to escape from molten plastic prior to being injected into a mold.

Chapter 9
BLACK SPECKS

Black specks can be defined as small dark particles or spots on the surface of an opaque part or within a transparent part.

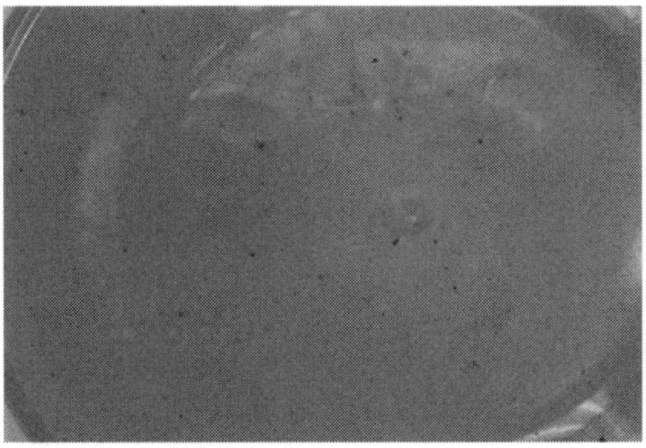

Figure 9-1 *Black specks.*

The following can cause black specks:
- Excessive residence time in barrel,
- trapped material,
- contamination in injection barrel,
- uncontrolled heater bands or thermocouples,
- damaged barrel or screw,
- oil leaks,
- sprue bushing nicked, rough, or not seating,
- improper venting,
- contamination from lubricants,
- mold too small for machine size,
- contaminated raw material,
- inconsistent process cycle.

MACHINE

Excessive Residence Time In Barrel

Explanation: Under the best conditions, a shot size should represent 50% of the capacity of the injection cylinder (barrel). This will result in processing the material for one cycle while preparing the material for the next cycle. Thus, a mold requiring a four-ounce shot should be run in a machine that has a barrel with an eight-ounce capacity. The more material left in the barrel between shots, the greater the likelihood of thermal degradation. This degradation is what causes the black specks.

Solution: Strive for a 50% shot-to-barrel ratio. This is ideal but can go as low as 20%, if the material is not too heat sensitive (like polypropylene) and up to 80% if the material is extremely heat sensitive (like PVC). It is not a good idea to empty the barrel for every shot because more time will be required to bring the next mass of material up to proper heat and degradation may occur.

Trapped Material

Explanation: If any molten resin is trapped along the flow path (most notably in the heating cylinder), it will stay there until it degrades. When this happens, the degraded material becomes carbonized, then chars and becomes brittle. At that point, it will flake away from the area of entrapment and enter the melt stream appearing as black specks or streaks.

Solution: Inspect the barrel liner, nozzle, non-return valve, and check ring for nicks, cracks, rough surfaces, peeled plating or stuck resin. Then, stone and polish as required, replace any damaged mechanisms, and inspect the main and secondary runners, as well as the sprue bushing, for nicks, rough surfaces or sharp corners. Round off sharp corners and radius corners where possible to minimize material trapping and shear points.

Contamination in Injection Barrel

Explanation: Any type of contamination in the injection barrel may be the cause of streaks, spots, and specks. It may be in the form of dust particles that fell from the ceiling into an open hopper, pellets from other materials, residual resin from an improper changeover, or even pieces of food that accidentally fell into a container of material ready to be placed in the hopper.

Solution: To remove this type of contamination it may be necessary to increase the temperature of the injection barrel and, using a purging material with a wide melt range, purge the contaminate(s) from the system.

Uncontrolled Heater Bands or Thermocouples

Explanation: Improperly sized or loose heater bands or thermocouples can cause localized degradation of the material by exposing it to extreme heat. They may be calling for more heat than normal due to malfunction or improper sizing. Even a heater band that is not working can be the cause of such overheating. The reason is that adjacent heater bands must increase heat to compensate for the nonworking band.

Solution: Check each heat zone to ensure that all heater bands are working properly, are properly controlled, properly sized, and are tight against the barrel. A conductive sealant should be used to ensure full contact with the barrel. Be sure to replace bands with the proper size, voltage, and wattage requirements as stated in the machine manual.

Damaged Barrel or Screw

Explanation: A cracked injection cylinder or pitted screw is a cause of material hang-up and degradation. Eventually this degraded material breaks loose and enters the melt stream, appearing as specks or streaks.

Solution: Inspect the injection unit for cracks and nicks in the walls. Sometimes damaged cylinder walls can be welded but it is usually more effective to replace the cylinder liner. Pitted screws must be welded, ground and replated, or replaced with new stock.

Oil Leaks

Explanation: Hydraulic components or fittings that are in the proximity of the injection cylinder may leak. This leakage may get into raw material storage containers and find its way into the material hopper. The oil will burn at the temperatures needed for molding and will degrade and char. This degraded material is a source for streaks and specks.

Solution: Eliminate all hydraulic leaks as soon as possible after they occur.

MOLD

Sprue Bushing is Nicked, Rough, or not Seating

Explanation: A damaged sprue may cause material to stick and be held in residence at elevated temperature until it degrades and decomposes. At that point, it will break loose and enter the melt stream as streaks or specks.

Solution: Inspect the internal surfaces of the sprue bushing. Remove any nicks or other imperfections. The tapered hole should be highly polished. Check the sprue bushing-to-nozzle seal with thin paper or bluing ink to ensure that the nozzle is centered to the bushing and that the hole and radius dimensions are compatible for the nozzle and the bushing.

Improper Venting

Explanation: Air is trapped in a closed mold and incoming molten plastic will compress this air until it auto-ignites. This burns the surrounding plastic and results in charred material in the form of spots and specks.

Solution: Vent the mold by grinding thin (0.0005"-0.002") pathways on the shutoff area of the cavity blocks. Vents should take up a minimum of 30% of the perimeter of the molded part. Vent the runner, too. Any air that is trapped in the runner will be pushed into the part.

Contamination from Lubricants

Explanation: Excessive use of mold release will clog vents. The trapped air cannot be evacuated and burns. Also, grease that is used for lubricating cams, slides, ejector pins, etc., can seep into the mold cavity and contaminate the molded part.

Solution: The remedy is to keep the mold as clean as possible and clean the vents if they become clogged. A white ash will be present if the vents are clogged. Also, make every effort to eliminate the use of external mold releases.

Mold Too Small for Machine Size

Explanation: If the mold is placed in too large of a machine, the chances are that the heating cylinder of that machine will be large enough to result in extensive residence time of the raw material in the heated cylinder. This will result in degraded material that will be injected into the mold causing streaks and specks.

Solution: Place all molds in properly sized machinery. A rule-of-thumb states that the machine should inject between 20% and 80% of its capacity every shot.

MATERIAL

Contaminated Raw Material

Explanation: The most common causes of black specks and streaks are molding compound contaminants. Such contamination is usually the result of dirty regrind, improperly cleaned hoppers or granulators, open or uncovered material containers, and poor quality virgin material supplied by the manufacturer.

Solution: This type of contamination can be minimized by dealing with high quality, reputable suppliers and by using good housekeeping practices. Properly trained material handlers will also help reduce contamination.

OPERATOR

Inconsistent Process Cycle

Explanation: It is possible that the machine operator is the cause of delayed or inconsistent cycles. This will result in excessive residence time of the material in the injection barrel. If such a condition exists, heat sensitive materials will degrade, resulting in black specks or streaks.

Solution: If at all possible, run the machine on automatic cycle, using the operator only to interrupt the cycle if an emergency occurs. Use a robot if an "operator" is really necessary. And, instruct all employees on the importance of maintaining consistent cycles.

Chapter 10

BLISTERS

Blisters can be defined as raised defects on the surface of a molded part caused by trapped gases in the part that could not escape before the surface began to "skin" during the molding process.

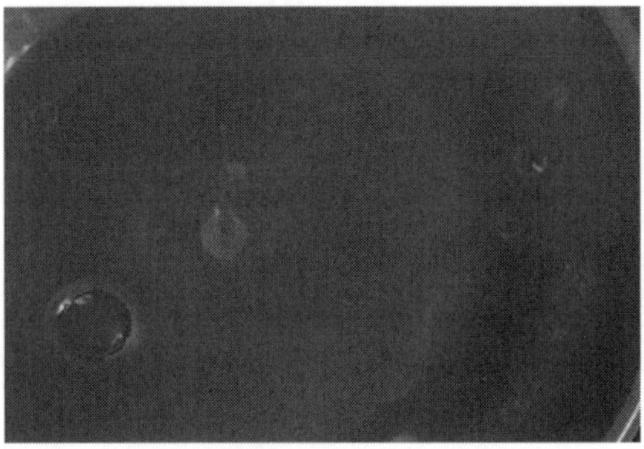

Figure 10-1 *Blisters.*

The following can cause blisters:

- Cycle time too short,
- screw rotation too high,
- low back pressure,
- injection speed too high,
- low mold temperature,
- high mold temperature,
- improper gate location,
- insufficient venting,
- use of regrind that is too coarse,
- use of highly volatile resins,
- excessive moisture,
- early gate opening.

MACHINE

Cycle Time Too Short

Explanation: In the process of minimizing overall cycle times to reduce manufacturing costs, many molders reduce the cooling portion of the cycle. This results in the surface skin of the molded part not being fully solidified when the part is ejected from the mold. Because of this, some of the gases that are formed during molding are allowed to expand against this soft skin. Blisters are formed because the gases are not constrained.

Solution: Although it may impact the manufacturing cost, the way to minimize blisters caused by short cooling time is to increase the "mold closed" portion of the cycle. Reducing mold temperatures may also help, but this practice may cause undue stress because of the increased injection pressure requirements. And, an increase of back pressure may help remove the gases before they enter the mold.

Screw Rotation Too High

Explanation: A screw rotation speed that is too high will tend to "whip" air into the molten plastic. This excessive air may not be drawn out of the material during the molding process and pockets of the air may be forced to the surface of the molded part, forming blisters. Excessive rotation speed also causes a shearing action that can cause plasticizing gases to develop. These gases can also cause blisters.

Solution: Slowing down the screw rotation will minimize the amount of air that is drawn into the material and will also minimize screw shear. Start at 100 RPM and adjust up or down as needed in 10-RPM increments.

Low Back Pressure

Explanation: The back pressure setting controls the density of the melt. A low setting results in a melt that is not dense enough to push out excessive gases. The gases may cause blisters.

Solution: Increasing the back pressure setting will make the melt denser and help remove gases and minimize trapped air volume.

Injection Speed Too High

Explanation: While it is usually best to inject at high speeds, too high a speed will cause a turbulence that traps air pockets. The air may not have a chance to escape through normal venting practices and may show up as surface blisters.

Solution: Reducing the injection speed will reduce the tendency for turbulence and trapped air pockets will not form.

MOLD

Low Mold Temperature

Explanation: As molten material enters the mold, the material starts to cool down immediately and a "skin" begins to form on the surface of the part being molded. If this skin forms too quickly, any air or gas that is in the material will not be allowed time enough to escape through proper venting methods. A mold that is too cold will cause this skin to form too early resulting in the air and gases being trapped and forming blisters.

Solution: Increasing the mold temperature will help to allow gases to escape by delaying the hardening of the surface skin.

High Mold Temperature

Explanation: Trapped air in a molten plastic will stay trapped inside the molded part if the skin forms properly before the part is ejected from the mold. A mold that is too hot does not allow that skin to form in time and the ejected part has a soft skin. Trapped air and gases can force their way through this skin and form blisters on the ejected parts surface.

Solution: Decrease the mold temperature to allow the skin to form in the right amount of time. This varies with various plastics. An increase in cooling time can also help form the required skin hardness but should only be used as a temporary fix as it will increase the cost of the molded part.

Improper Gate Location

Explanation: As material travels through the gate and enters the cavity it seeks the path of least resistance. An improper gate location can cause the material to take an improper path and not properly push the trapped air out in front of it.

Solution: Consideration of material flow paths and vent locations at the mold design stage will minimize trapped air blister problems on new molds. Existing molds may require relocating the gate. Gates should normally be located in the thickest section of the part.

Insufficient Venting

Explanation: The correct size, location, and shape of vents should be considered and analyzed in the mold design stages. Inadequate venting will not allow trapped and gases to escape from a mold and this may result in blisters, burns, or other defects.

Solution: Venting is a very important part of the hole molding process. The parting line perimeter of the cavity should contain vents equal to 30% of that perimeter. Another rule-of-thumb is to place a vent at every inch along the parting line perimeter. And, the runner should be vented.

MATERIAL

Use Of Regrind That Is Too Coarse

Explanation: Using excessively coarse regrind increases the amount (volume) of air that gets trapped in the melt because the coarse, uneven particles of regrind create "pockets" of air between them and the smaller, consistently sized particles of virgin material. These pockets of air get pushed into the molded part and may result in surface blisters.

Solution: One remedy is to use a finer gauge screen in the granulator. This will produce smaller particle sizes. Another remedy is to limit the amount of regrind that is used to less than 5%. Also, an increase in back pressure may help blend out the trapped air. And, the final solution is to use only virgin material.

Use of Highly Volatile Resins

Explanation: Some molding materials (such as liquid crystal polymer) release a large amount of volatile gases during the plasticizing portion of the molding process. These gases need a chance to escape from the injection barrel before being injected into the mold.

Solution: Back pressure control can be used to accommodate this condition, and some success has been achieved through utilization of vented barrels on the machine itself.

Excessive Moisture

Explanation: Improperly dried or stored molding material will contain excessive moisture because all plastics either have a tendency to absorb moisture from the atmosphere or hold moisture that has accumulated through condensation or spillage. When processed, this moisture turns to steam in the melt flow and will form pockets of trapped gas that may show up as blisters in the molded part.

Solution: Properly dry the material before using and store it correctly to minimize future absorption or accumulation of moisture. Most materials need to be dried to a dew point reading of between -20 and -40 degrees F. This equates to a level of less than 0.10% by weight. And remember that the material must be used within two hours of drying or moisture can accumulate again. Even regrind must be dried before using if it is allowed to stand for more than two hours.

OPERATOR

Early Gate Opening

Explanation: It is possible that the machine operator can cause blisters by opening the safety gate too soon. Depending on the age and type of molding machine, this can cause the mold to open before a hard skin has had time to form on the molded part. Trapped gases will be allowed to expand and form blisters on the surface of the part.

Solution: If possible, run the machine on automatic cycle, using the operator only to interrupt the cycle if an emergency occurs. Use a robot if an "operator" is really necessary. And, instruct all employees on the importance of maintaining consistent cycles.

DEFECTS

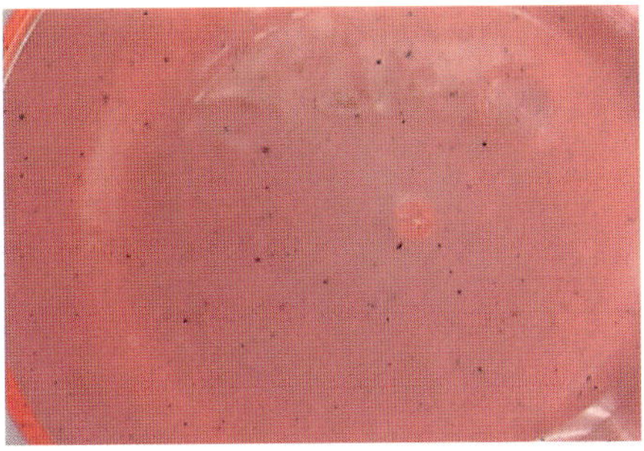

Black Specks

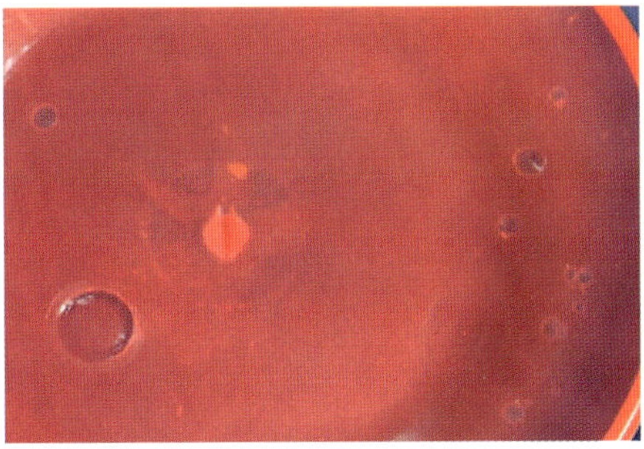

Blisters

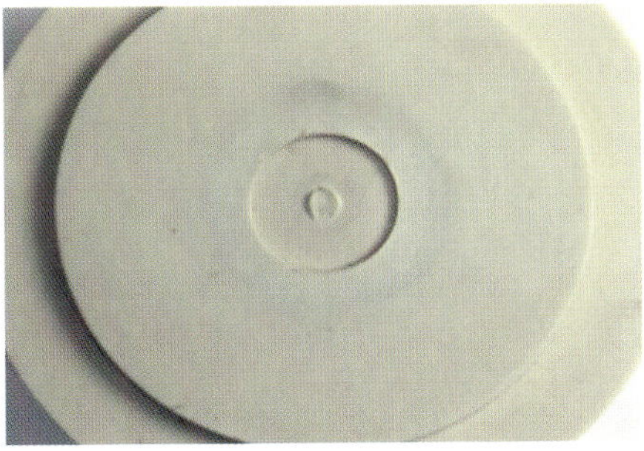

Blush

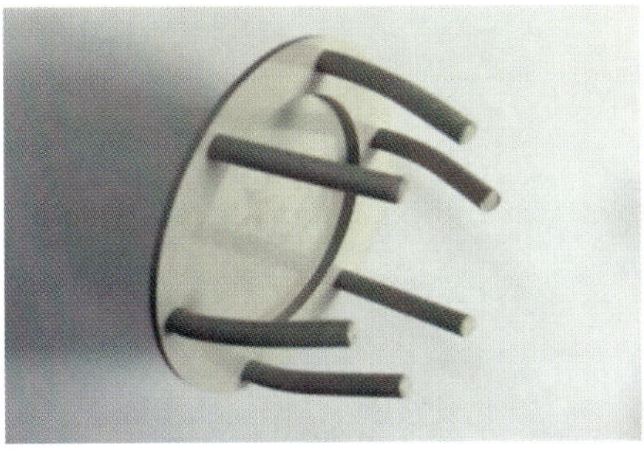

Bowing

Brittleness

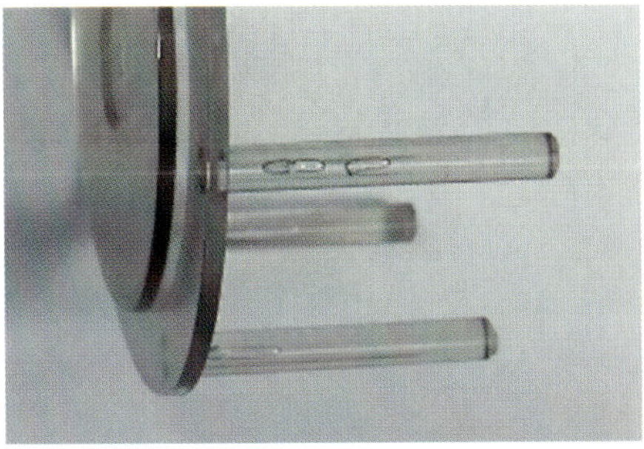

Bubbles

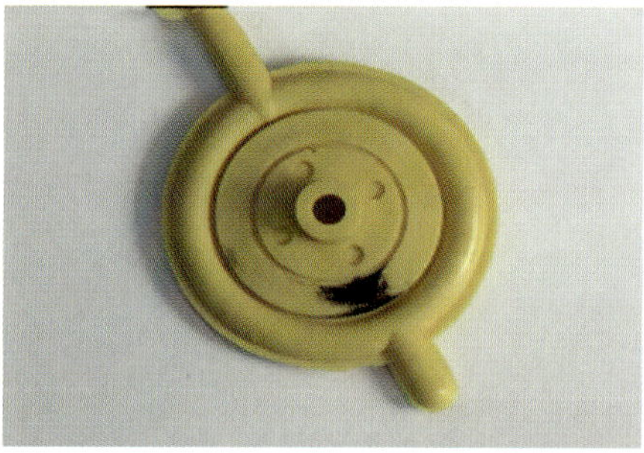

Burn Marks

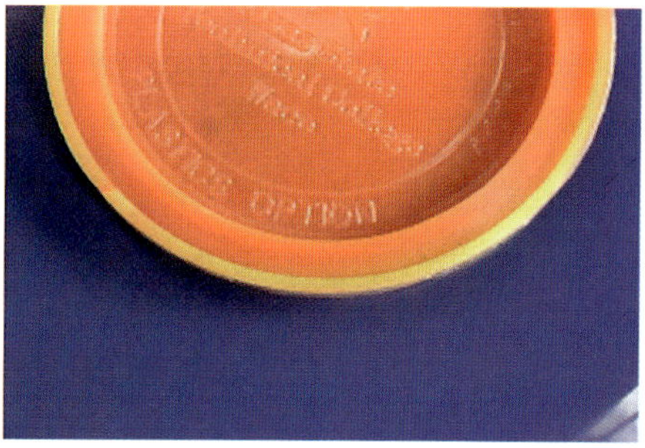

Clear Spots

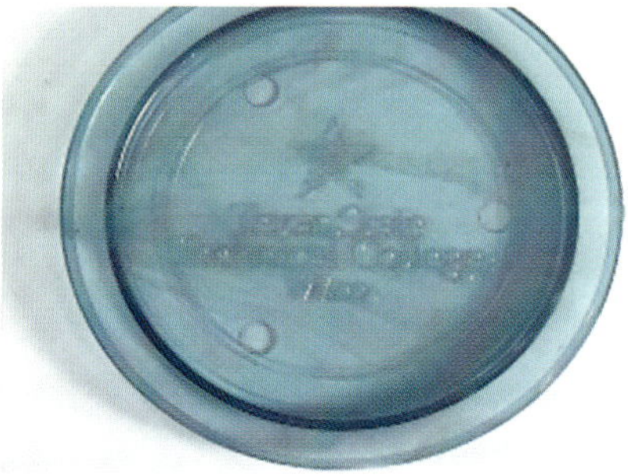

Cloudy Appearance

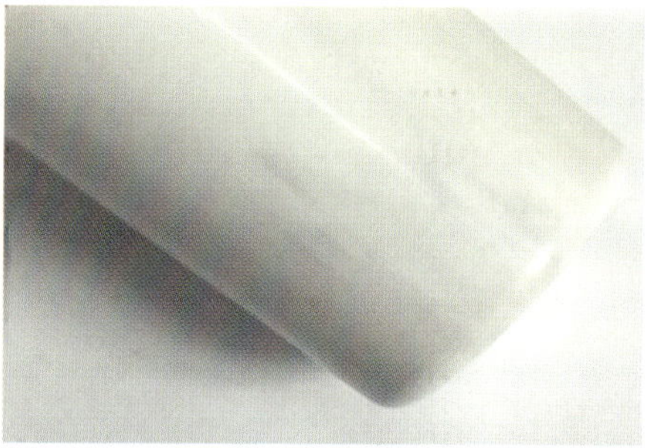

Contamination

Cracking

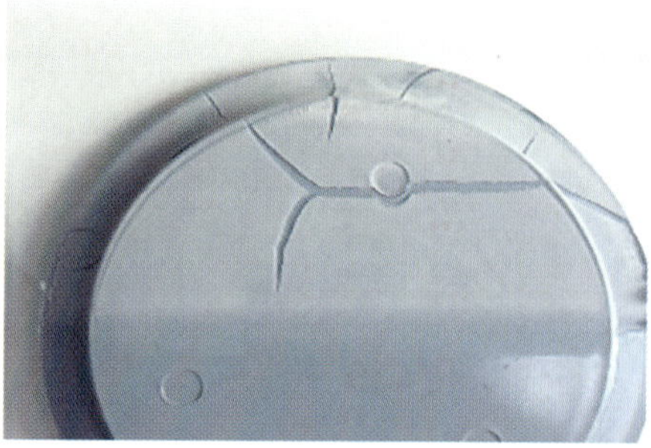

Crazing

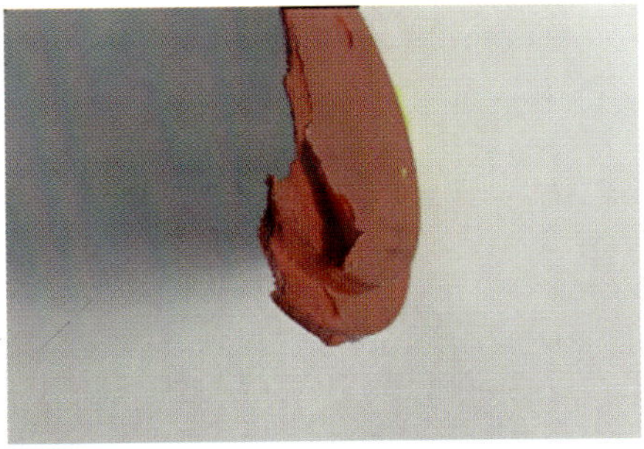

Delamination

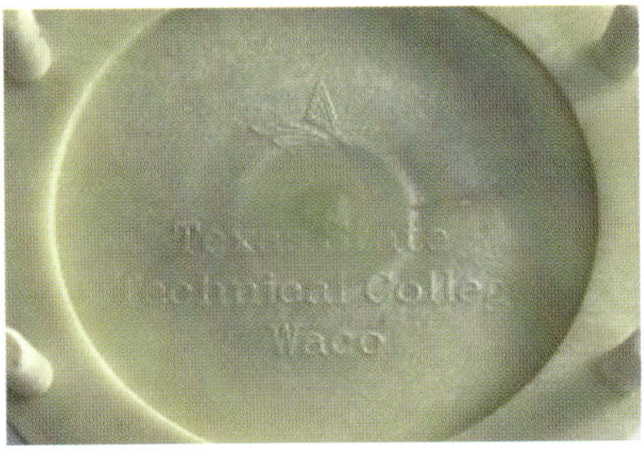

Discoloration

Flash

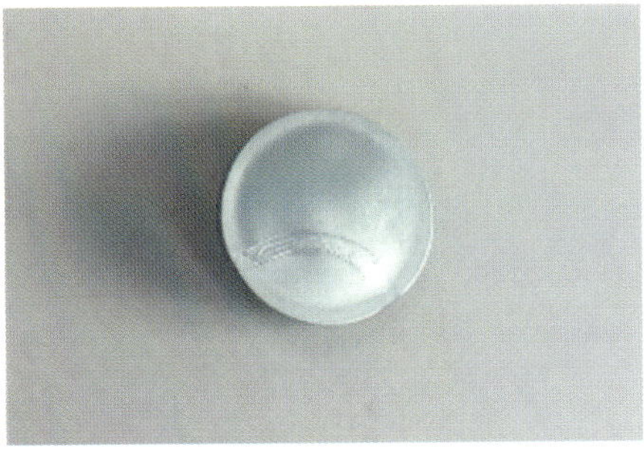

Flow Lines

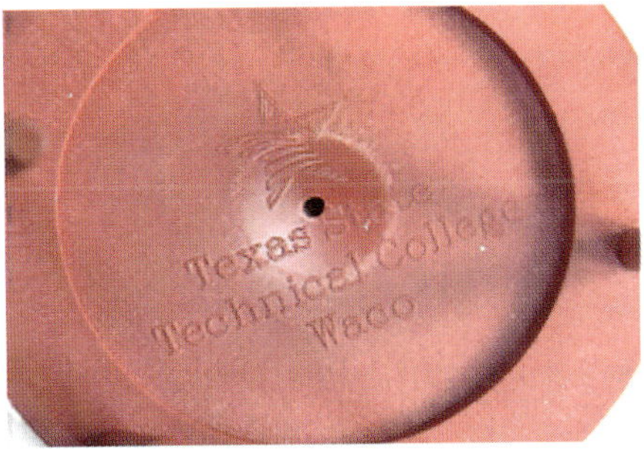

Gloss

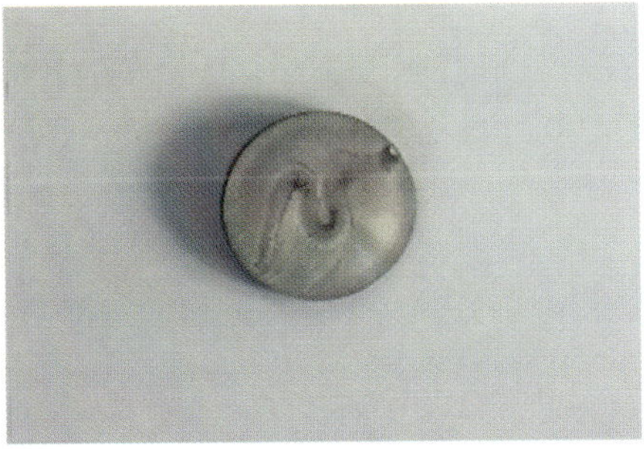

Jetting

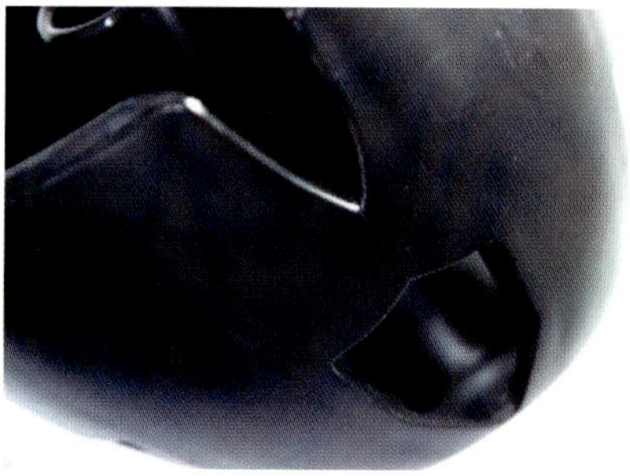

Knit Lines

NonFill

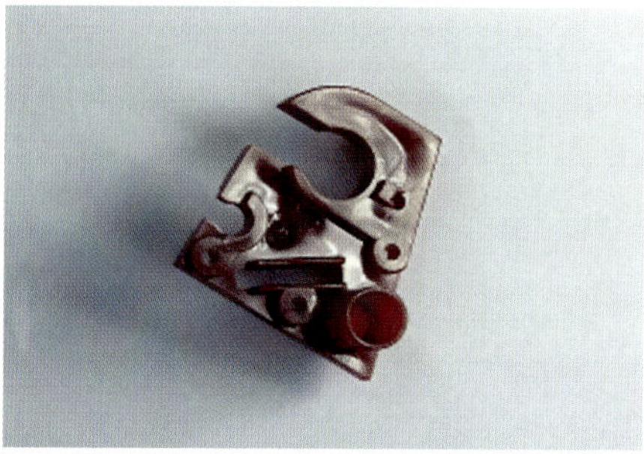

Shrinkage

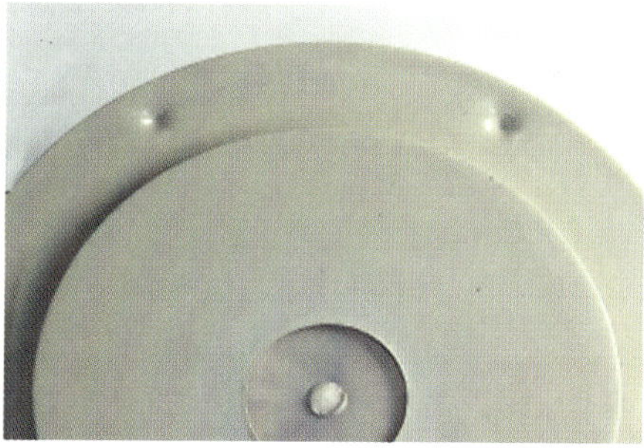

Sink Marks

Splay

Warpage

Chapter 11
BLUSH

Blush can be defined as a clouded discoloration normally found at gate locations, but they can be found anywhere on the part surface. These visually defective areas have very weak physical properties due to loose molecular structure.

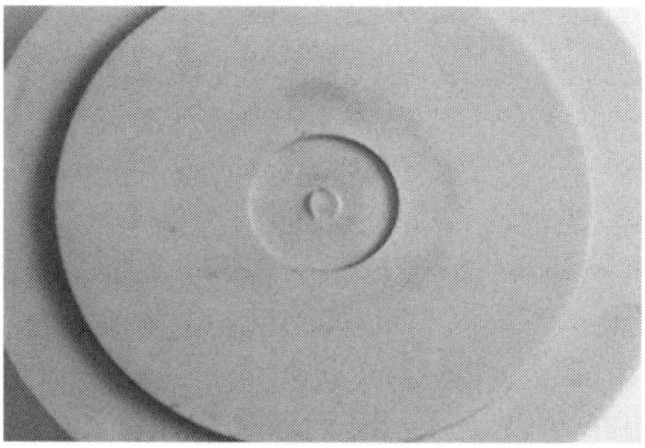

Figure 11-1 *Blush.*

The following can cause blush:
- Excessive injection fill speed,
- melt temperature too high or too low,
- low injection pressure,
- nozzle diameter too small,
- low nozzle temperature,
- low mold temperature,
- improper venting,
- small sprue bushing diameter,
- improper gate location,
- sharp corners,
- excessive moisture,
- inconsistent process cycle.

MACHINE

Excessive Injection Fill Speed

Explanation: The speed and pressure of the melt as it enters the mold determine both density and consistency of melt in packing the mold. If the fill is too fast, the material tends to "slip" over the surface and will "skin" over before the rest of the material solidifies. The slipped skin area does not faithfully reproduce the mold steel surface, as does the material in other areas, because it has not been packed tightly against the steel.

Solution: One solution is to adjust the fill speed rate until the optimum has been achieved. This will help eliminate blushing.

Melt Temperature Too High or Too Low

Explanation: Although this may sound contradictory, either condition might cause blushing. If the injection barrel heat is too high, the material will flow too quickly, resulting in slippage of the surface skin, as mentioned above. If the barrel heat is too low, the material may solidify before full packing occurs and the plastic will not be pushed against the mold steel, especially in the gate area because that is the last area to pack.

Solution: Melt temperature must be adjusted to the optimum for a specific material and specific product design.

Low Injection Pressure

Explanation: The plastic material must be injected into the mold in such a way as to cause proper filling and packing while maintaining consistent solidification of the melt. Injection pressure is one of the main control variables of the machine and must be high enough to pack the plastic molecules against the steel of the mold while the plastic cools. Low pressure will not achieve this packing and the material will appear dull in local areas that do not have enough pressure.

Solution: Increasing the injection pressure forces the material against the mold surface, producing a truer finish that replicates the steel finish.

Nozzle Diameter Too Small

Explanation: The nozzle diameter controls the time during which the material fills the mold. If the diameter is too small, the material may begin to solidify before the mold is filled. Then, packing cannot occur because the material is already rigid. Blush will occur because the plastic has not been forced against the steel surface.

Solution: Enlarging the nozzle diameter will minimize the condition. The nozzle tip is interchangeable and a tip with the opening the same as, or slightly smaller than, the sprue bushing opening is recommended.

Low Nozzle Temperature

Explanation: A nozzle that has a low temperature will cause the material going through it to cool off too soon and not be allowed to pack out the mold. The non-packed molecules will form blush because they cannot replicate the steel finish.

Solution: Increasing the mold temperature will allow the material to flow easier, and for a longer time, thereby packing the mold and replicating the steel finish. Normally, the nozzle temperature should be set the same as, or 10 degrees hotter than, the front zone of the barrel.

MOLD

Low Mold Temperature

Explanation: A low mold temperature may cause the molten material to slow down and solidify before the mold is packed out. This will cause dull areas where the plastic was not forced against the steel finish.

Solution: Increasing the mold temperature allows the material to flow farther and pack properly. The material temperature could also be raised to accomplish the same effect.

Improper Venting

Explanation: Trapped air can cause blushing if the air is trapped in an area that does not compress the air enough to ignite it. The air takes up space where the plastic should be, so the plastic is not forced against the steel finish.

Solution: Vent the mold by grinding thin (0.0005"-0.002") pathways on the shutoff area of the cavity blocks. Vents should take up approximately 30% of the perimeter of the molded part. Add vents in local areas that show blush. Vent the runner, too. Any air that is trapped in the runner will be pushed into the part.

Small Sprue Bushing Diameter

Explanation: A small sprue-bushing diameter will keep the material from packing because the small opening reduces the ability of the plastic to flow far enough to fill the mold. An unpacked mold will cause blushing where the material is not forced against the mold steel.

Solution: Size the sprue bushing major diameter so its cross-sectional area is equal to (or greater than) the sum of the cross-sectional area of all the runners leading from it. Then, taper the sprue diameter to match the nozzle. That will ensure proper pressure drop adjustments to pack the mold.

Improper Gate Location

Explanation: If a mold is gated such that the thinnest areas fill first, those areas will begin to solidify before the thicker areas are packed. Blush will form in the thicker areas because there is no pressure left to pack the plastic against the steel surface.

Solution: Make sure the gate is located so the thicker sections fill first. The material should flow from thick section to thin section. That ensures equal packing of all areas.

Sharp Corners

Explanation: If the product design contains sharp corners, the material tends to slip by those corners without fill them in. The corners are not packed with material and blush occurs due to that non-packing.

Solution: Radius all sharp corners, especially in the gate area, as that is the last place to pack. Sharp corners should not be allowed on any molded part.

MATERIAL

Excessive Moisture

Explanation: In some cases, excessive moisture in a melt will accumulate at the gate area. The reason for this is that the gate area is the last place the pressure builds up. Moisture trapped in other areas may be forced into the gate area due to this pressure buildup. The gate area will appear dull due to the moisture that gets screened out. Usually, this type of blush is accompanied by splay.

Solution: Dry the material to the supplier's recommendations and make sure it is used within two hours of that drying activity.

OPERATOR

Inconsistent Process Cycle

Explanation: It is possible that the machine operator is the cause of delayed or inconsistent cycles. This will result in excessive residence time of the material in the injection barrel. If such a condition exists, materials may fill at a faster speed and cause slippage as explained earlier. Slippage causes blush.

Solution: If possible, run the machine on automatic cycle, using the operator only to interrupt the cycle if an emergency occurs. Use a robot if an "operator" is really necessary. And, instruct all employees on the importance of maintaining consistent cycles.

Chapter 12
BOWING

Bowing can be defined as a condition of being unintentionally bent into the shape of a bow. This shape can be either convex or concave and is usually caused by a differential in shrinkage rates from one face of a part to an opposite face, but can also be caused by mechanical distortion.

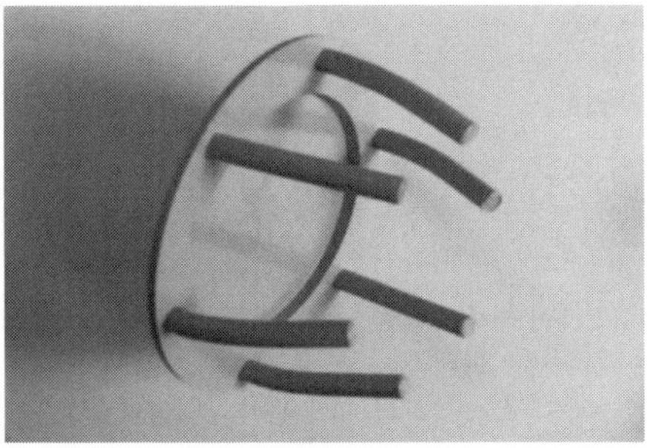

Figure 12-1 *Bowing.*

The following can cause bowing:
- Clamp opens too quickly,
- ejector system not level,
- cooling time too short,
- inadequate packing of molecules,
- temperature too low,
- inconsistent mold temperature,
- improper gate location,
- melt index too low,
- parts mishandled after ejection.

MACHINE

Clamp Opens Too Quickly

Explanation: To produce more parts in a given amount of time, molders sometimes cause the clamp to open before the part has cooled enough in the mold. The part may not have shrunken enough on to the "B" half of the mold and wants to still stay on the "A" half. This results in a temporary tug-of-war until the part snaps back onto the "B" half as the mold continues to open. The snapping action can permanently distort the part.

Solution: Make sure the mold is closed for the proper amount of time, and allow it to open slowly for the first 1/2" or so. That should be enough to allow the part to shrink to the "B" half. Then, the mold can continue opening more quickly.

Ejector System Not Level

Explanation: Due to wear, lack of lubrication, age, improper settings, or a variety of other reasons, the machines ejector system may go out of alignment. The system may come forward at a jaunted angle and force the molds ejector to come forward at the same angle. The molded part may be ejected in a cocked manner and may take a permanent set in that distorted condition.

Solution: Analyze the entire ejection system of the machine. Readjust and align anything that is out of adjustment and make sure locks are used to prevent future movement. Lubricate the system regularly and make sure you are using guided ejector systems in the molds if possible.

Cooling Time Too Short

Explanation: Plastic material must be held (constrained) in the mold for a long enough period of time to allow a solid skin to form on the part. If cooling time is too short, the material will continue to move as it cools outside the press and, with nothing to constrain it, will distort until it stabilizes.

Solution: Increase the cooling portion time of the cycle to ensure proper skin solidification of the plastic material.

Inadequate Packing Of Molecules

Explanation: The amount of material being injected into the cavity determines whether or not adequate pressures can build up against that material and pack the molecules together to hold them in place during the cooling portion of the cycle. Also, the amount of time used for the injection "hold" phase of the cycle determines how much packing is achieved. Too little hold time will result in the material partially being sucked back out of the cavity, which relaxes the total pressure being held against the material remaining in the cavity. Too much hold time may result in the material solidifying without being able to properly shrink slightly away from the cavity walls. This is called "*over-packing*" and the parts will be distorted.

Solution: Adjust the material feed so that a "cushion" of material is present at the end of the injection stroke. The standard requirement is 1/8" but it can go to 1/4" for most materials. Adjust the injection hold time to stay engaged until the gate freezes. Then, the screw can be retracted without affecting material that is in the cavity.

MOLD

Temperature Too Low

Explanation: Some materials (such as polyesters) require mold temperatures that are above the boiling point of water (212 degrees F). All materials require mold temperatures high enough to sustain proper flow and packing in the mold. Mold temperatures that are too low result in inadequate filling and uneven packing. Uneven packing results in uneven shrinkage and this will cause bowing and distortion.

Solution: Raise the mold temperature to that recommended by the material supplier for the specific material being molded. This may require the use of an oil heater, or cartridge heaters placed in the mold, if the requirement is higher than 200 degrees F.

Inconsistent Mold Temperature

Explanation: A molded part must be cooled in a consistent manner. Uneven cooling will result in uneven shrinkage and that can cause bowing. There should be no hot spots or cold areas in the mold around the molded plastic. In fact, there should be no more than a 10 degree F difference between any two points on the cavity surfaces when measured with a surface pyrometer. A difference of more than 10 degrees will result in uneven cooling. Bowing can also occur if one half of the mold is hotter than the other. The part will want to stay to the hotter half and this can cause bowing as the mold opens.

Solution: First, make sure that each mold half has its own temperature control system. Then, check for hot or cold spots in the mold using a surface pyrometer. The only areas of concern are where the plastic will be molded (including the runner). The rest of the mold does not matter. If you find more than a 10 degree F difference, adjustments must be made to the cooling system or pattern by adding cooling lines, bubblers, or other methods.

Improper Gate Location

Explanation: A gate should be designed and located such that the material flow will be consistent in all directions. The material should flow equally through the cavity both in speed and volume. And, in multiple cavity molds each cavity should finish filling at the exact same instant. This results in even fill, even packing, and even shrinkage and is known as a *balanced runner system*. If the system is unbalanced, or improperly located, uneven pressures will cause uneven shrinkage and bowing will occur.

Solution: Always locate the gate in the thickest section of the part and allow material to flow from thick to thin. Rectangular parts should have the material flow across the width. Circular parts should be centrally gated. This will result in fast filling and even shrinkage.

MATERIAL

Melt Index Too Low

Explanation: Every material is available in a range of Melt Index values. This number (average is 14) indicates the flow-ability of a material. The higher the number, the easier the material flows. If the material is purchased at the low end of the range it will be stiffer than at the high end of the range. Stiffer materials are more difficult to push and will require higher injection pressures to fill the mold. Unfortunately, these high pressures tend to over-pack the material and less shrinkage occurs. When the mold opens the parts may distort.

Solution: Use a Melt Index (MI) value that is mid-range to start with. You can always request a higher or lower value later. But, be practical. The value must have a tolerance applied. For example, if you determine that a 14 MI is the right value, understand the tolerance factor will provide MIs of at least 13 to 15.

OPERATOR

Parts Mishandled After Ejection

Explanation: Even if parts are molded properly and have no uneven shrinkage, they are still hot enough to be distorted by the operator if they are mishandled after they are ejected. Shoving the parts together into a box, or laying them so they are not flat, may cause the parts to distort and hold the distorted shape as they cool further.

Solution: Instruct the operator on the proper method of handling the parts. Parts should be allowed to air cool for a minimum of six cycles before being packaged, and then they should be packaged loosely.

Chapter 13
BRITTLENESS

Brittleness can be defined as the tendency of a molded plastic part to break or crack under conditions in which it would not normally do so. At times the part may also shatter.

Figure 13-1 *Brittleness.*

The following can cause brittleness:
- Improper screw design,
- short cycle time,
- excessive back pressure, screw rpm, or injection speed,
- excessive nozzle temperature,
- low injection pressure,
- gate and/or runner restrictions,
- condensation or leaks,
- resin too cold,
- excessive moisture,
- degraded resin,
- inconsistent process cycle.

MACHINE

Improper Screw Design

Explanation: An injection screw with too low of a compression ratio will not properly mix and melt the material. This results in weak bonding of individual plastic resin molecules. The weak molecular bonds cause the molded part to be brittle.

Solution: Use an injection screw with the proper compression ratio. The material supplier is the best source for this information. While the general purpose screw that comes with the machine is adequate for many situations, specific screw designs are available for almost any specific material.

Short Cycle Time

Explanation: The overall machine cycle time may be so short that the material does not have enough residence time in the injection barrel to melt to the proper consistency. This results in material that does not get thoroughly mixed and melted. A weak bond occurs between molecules and the part is brittle due to that weakness.

Solution: Increasing the cycle time will allow longer residence time for the material, and it will heat and melt better. However, this will add to the cost of molding, so alternative actions should be taken first. For example, increasing the barrel temperature, increasing screw RPM and increasing back pressure will have the same effect. Be cautious, though. Too great an increase in these areas will cause degradation of the material (see the next paragraph).

Excessive Back Pressure, Screw RPM, or Injection Speed

Explanation: Increasing back pressure, screw speed, and/or injection fill rate beyond recommendations for specific materials will result in thermal degradation of the plastic through increased shear heat. The overheated resin forms a weak molecular bond resulting in brittleness.

Solution: Follow the material supplier's recommendations concerning these parameters and do not go beyond or above their suggested values.

Excessive Nozzle Temperature

Explanation: A nozzle that is too hot will overheat the material passing through. This may cause a separation of molecules due to localized thermal degradation and will result in a lack of proper molecular bonding, which will cause brittleness.

Solution: A reduction of nozzle temperature to that of the front zone is a good starting point. Gradually adjust the nozzle temperature, if necessary, to optimum conditions. Normally that optimum temperature would be 10 degrees F above the front zone, but it may vary with specific materials.

Low Injection Pressure

Explanation: Low injection pressure may result in the formation of weld line areas in the molded part. This is caused by a non-filling condition because the plastic solidifies before it can fully pack the cavity. The result is a weak area that tends to appear brittle and may even exhibit signs of cracking.

Solution: Increasing the injection pressure helps to force the weld line areas together, minimizing the tendency to crack or appear brittle.

MOLD

Gate and/or Runner Restrictions

Explanation: Gates and runners that are too small will cause restrictions to the flow of molten plastic. These restrictions cause the material to heat up due to shearing friction and the material will thermally degrade. This results in weak molecular bonding and causes the molded parts to be brittle.

Solution: Examine the gates and runners and, if possible, perform a computer simulation to determine the optimum size and shape of runner and gate for the specific parts(s) being molded. Remember that too large a gate and runner is just as detrimental as too small a gate and runner. Follow the material supplier's recommendations.

Condensation or Leaks

Explanation: Condensation being formed on the surface of a cold mold in humid conditions will cause moisture to form and be carried into the cavity by the molten plastic. This moisture forms a barrier to localized molecular bonding and this will result in brittleness. Also, if there is a water leak in the cavity due to cracks (no matter how small) moisture will be picked up by the incoming plastic and cause brittleness.

Solution: Raising the mold temperature will eliminate "sweating" (condensation) on the mold surfaces. Check for crack conditions in the mold cavities. Sometimes, leaking "O" ring seals will be the source of water leakage.

MATERIAL

Resin Too Cold

Explanation: A cold resin will result in a poorly mixed and blended melt. There may even be "clumps" of unmelted resin that are not bonded to other "clumps". This results in a weak area lacking proper molecular bonding with localized brittleness or cracking.

Solution: Increasing the barrel temperature, and/or back pressure, will help soften and homogenize the plastic and result in stronger molecular bonds.

Excessive Moisture

Explanation: Excessive moisture is one of the most frequent causes of brittleness. Moisture causes brittleness because the water droplets actually turn to steam when heated in the injection unit, and this steam explodes throughout the plastic, interfering with molecular bonding, causing voided areas between molecules. This causes those areas to be extremely weak and brittle. The voided areas easily break apart when exposed to any mechanical forces.

Solution: Although it is commonly understood that non-hygroscopic materials do not require drying, do not take chances. Dry all materials. It may be that fillers used in the material **are** hygroscopic and they will absorb moisture. Every plastic material requires specific drying conditions, and each material should be dried according to the material supplier's recommendations. The desired moisture content is between 1/10th of 1 percent and 1/20th of 1 percent by weight. This means the dry air being used to take moisture from the material should have a dew point of -20 to -40 degrees F.

Degraded Resin

Explanation: It is certainly possible to degrade a material by overheating it. This can be done in a variety of ways, including too high a barrel temperature, too long a residence time in the barrel, too much restriction in the runner system or gate, and too fast of an injection fill rate. Molecules of the plastic will break their bonds when overheated and the material may even char or burn locally. With the broken bonds, the material is weak and brittle.

Solution: Reduce the temperature of the plastic. Shortening the overall cycle time, reducing residence time, reducing barrel temperatures, or moving the mold to another machine with a smaller barrel can accomplish this. Runners and gates also should be analyzed to improve any shearing situation.

OPERATOR

Inconsistent Process Cycle

Explanation: It is possible that the machine operator is the cause of delayed or inconsistent cycles. This will result in excessive residence time of the material in the injection barrel. If such a condition exists, heat sensitive materials will degrade due to erratic heating in the barrel, resulting in brittleness.

Solution: If possible, run the machine on automatic cycle, using the operator only to interrupt the cycle if an emergency occurs. Use a robot if an "operator" is really necessary. And, instruct all employees on the importance of maintaining consistent cycles.

Chapter 14
BUBBLES (VOIDS)

Bubbles can be defined as a voided area trapped within a molded plastic part. It differs from a blister in that there is no surface protrusion with a bubble. Bubbles are usually caused by trapped gases or air pockets, but can also be caused by differential shrinking.

Figure 14-1 *Bubbles.*

The following can cause bubbles:
- Low injection pressure or hold time,
- insufficient material,
- improper injection temperature profile,
- excessive injection speed,
- insufficient back pressure,
- improper venting,
- section thickness too great,
- improper runners or gates,
- low mold temperature,
- excessive moisture,
- inconsistent process cycle.

MACHINE

Low Injection Pressure or Hold Time

Explanation: If injection pressure or hold time are too low, the molten material is not forced into the mold cavity and trapped gases and air will form voids because the gases will not be forced out of the mold through vent paths.

Solution: Increase the injection pressure and/or the hold time to help force the gases out as the plastic is pushed into the cavity.

Insufficient Material

Explanation: Too little material feed will have the same effect as low injection pressure. The material will not be forced into the cavity and gases will be trapped, forming voids due to a lack of molecular packing.

Solution: It is important to establish a feed setting that allows a 1/8" to 1/4" cushion of material at the end of the injection stroke. Without this cushion, there is no material against which holding pressure can be applied to force material into the cavity.

Improper Injection Temperature Profile

Explanation: The injection temperature profile addresses four heating zones of the injection barrel. These are commonly known as rear, center, front, and nozzle. The rear is also know as the feed zone, the center is known as the transition zone, and the front is known as the metering zone. The purpose of the feed zone (rear) is to start the material through the heating process. The heat is kept lower at this point but high enough to begin softening the plastic. The transition zone (center) heats the plastic higher and begins to compress it, squeezing out the trapped gases. In the metering zone (front) the material is brought up to the final, ideal temperature and is further compressed and sheared, which also introduces more heat. In the nozzle zone the material is simply kept at the upper temperature as it is injected into the mold. Any imbalance in the temperature values of these zones may result in plastic particles that are not properly melted at the right time. This will not allow gases to escape and voided areas (bubbles) will appear.

Solution: Maintain a proper temperature profile. This is readily obtained from the resin supplier, but a rule-of-thumb sets the temperature controls at increments of 50 to 100 degrees F from rear to front, and the nozzle at the same temperature as the front zone. An air shot from the nozzle should produce a bubble-free stream of plastic that has the approximate consistency of warm honey. Remember that the temperature control settings are not the same as the actual temperature of the plastic. They are usually 50 to 100 degrees higher than the actual plastic temperature to accommodate the rapid travel of material through the barrel.

Excessive Injection Speed

Explanation: The injection speed determines how fast the material is injected into the mold. If it is too slow, the material tends to cool off and solidify before the mold is fills, which results in a short shot. If it is too fast, the material tends to tumble and become turbulent, which traps air and gases in the resin. These gases then show up as bubbles because they were not able to reach the vented areas of the mold.

Solution: Start with the supplier's recommendations for injection fill speed. Adjust up or down according to the results. If bubbles appear, slow down the rate. If short shots appear, speed up the rate.

Insufficient Back Pressure

Explanation: Back pressure is used to help mix the material and homogenize it. It also helps remove trapped air and densifies the melt. If back pressure is insufficient the gases and trapped air are not allowed to escape and remain in the plastic melt as bubbles that can be molded into the finished part.

Solution: Increase the back pressure. Most materials will benefit from a back pressure that is approximately 50 psi. But, some materials require higher settings: in some cases up to 300 psi. However, be cautious, because too high a back pressure will degrade any material. The material supplier is the best source of information regarding proper back pressure settings.

MOLD

Improper Venting

Explanation: Most molds do not have adequate venting. Usually the moldmaker elects to "wait and see" where the venting needs to be located and then assigns an arbitrary size. While size is not necessarily as important as location, there is a tendency to use a minimum number of oversized vents rather than an adequate number of properly sized vents. If improper venting is used (or no venting), any trapped air or generated gases cannot escape. This will result in voids, bubbles, shorts, and burns.

Solution: Vent the mold even before the first shot is taken by grinding thin (0.0005"-0.002") pathways on the shutoff area of the cavity blocks. Vents should take up approximately 30% of the perimeter of the molded part. Vent the runner, too. Any air that is trapped in the runner will be pushed into the part. Another rule-of-thumb is to place a vent at every 1-inch dimension around the perimeter of the cavity. You cannot have too many vents.

Section Thickness Too Great

Explanation: Most plastic parts are not of one continuous wall thickness. There is usually a need to change the wall thickness for such reasons as additional strength. Unfortunately, when that happens, there is a pressure loss in the thicker section as the molten material shrinks more there as it solidifies. The material pulls away from the cavity wall leaving a voided area. If the void is captured below the part surface, the void will appear as a bubble.

Solution: A good rule-of-thumb is that any wall thickness should not exceed any other wall thickness by more than 25%. There will be little tendency for bubbles at that ratio. Metal inserts can be used to core out sections that do not meet that ratio, or "overflow" wells might be used to move the voided area off the primary part surface. However, the overflow would then need to be removed from the molded part.

Improper Runners or Gates

Explanation: Runners or gates that are too small will restrict the molten material in the flow pattern and may cause non-packed parts. If gates are placed to flow material from a thin section into a thicker section, the restricted flow in the thin section will keep the thicker section from packing. Both of these conditions can result in a loss of filling pressure and cause sinks to evolve in the molded part. These sinks can take the shape of bubbles and voids if they are trapped within the part rather than on the surface.

Solution: Gates should be of a depth that is equal to at least 50% of the wall they are placed at and should always be located to flow material from the thickest section to the thinnest. Runner diameters should be adequate to avoid a pressure drop as the material fills. Thus, the farther the travel, the larger the initial runner diameter should be. Gates and runners should be machined in the mold to be "steel safe" so they can be increased by removing metal. It is a good practice to place gates and runners in individual inserts so they can be easily replaced and/or reworked.

Low Mold Temperature

Explanation: A mold that is too cold for a specific resin or product design will not allow the molten material to fill and pack all of the mold properly before the resin starts to solidify. Any air or gases present in the plastic at the time will be trapped under the surface as bubbles.

Solution: Raise the mold temperature in increments of 10 degrees F until the bubbles disappear. Allow 10 cycles for each 10-degree adjustment (up or down) for the mold temperature to stabilize.

MATERIAL

Excessive Moisture

Explanation: Excessive moisture is one of the most frequent causes of bubbles. Moisture causes bubbles because the water droplets actually turn to pockets of steam when heated in the injection unit, causing voided areas between molecules. If the voided areas are trapped beneath the surface of the part they appear as bubbles.

Solution: Although it is commonly understood that non-hygroscopic materials do not require drying, do not take chances. Dry all materials. It may be that fillers used in the material are hygroscopic and they will absorb moisture. Every plastic material requires specific drying conditions. And each material should be dried according to the material suppliers recommendations. The desired moisture content is between 1/10th of 1 percent and 1/20th of 1 percent by weight. This means the dry air being used to take moisture from the material should have a dew point of -20 to -40 degrees F.

OPERATOR

Inconsistent Process Cycle

Explanation: It is possible that the machine operator is the cause of delayed or inconsistent cycles. This will result in excessive residence time of the material in the injection barrel. If such a condition exists, materials may flow more easily and be injected too quickly, resulting in trapped air and gases being held in the resin and not being vented as required. The gases will form bubbles if held under the molded part surface.

Solution: If possible, run the machine on automatic cycle, using the operator only to interrupt the cycle if an emergency occurs. Use a robot if an "operator" is really necessary. And, instruct all employees on the importance of maintaining consistent cycles.

Chapter 15
BURN MARKS

Burn marks can be defined as small dark brown or black discolorations on the surface of a molded part, usually found at the end of the material flow path or in blind pockets.

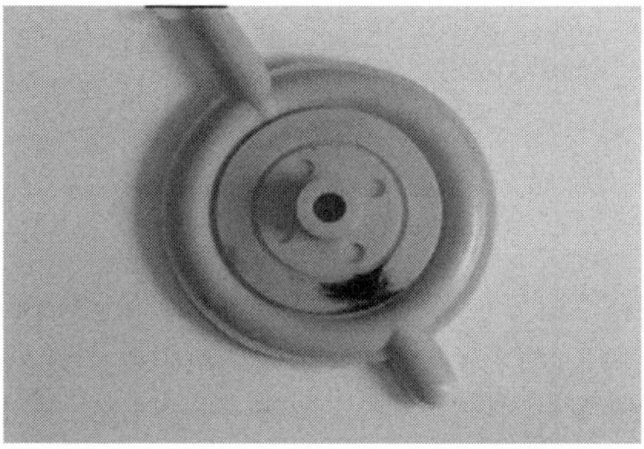

Figure 15-1 *Burn marks.*

The following can cause burn marks:
- Excessive injection speed or pressure,
- excessive back pressure,
- high screw rotation speed,
- improper compression ratio of screw,
- faulty temperature controllers and heater bands,
- excessive barrel temperatures,
- improper sprue bushing-to-nozzle sizing,
- improper venting,
- undersized gates,
- excessive use of regrind,
- inconsistent process cycle.

MACHINE

Excessive Injection Speed or Pressure

Explanation: Injection speed and pressure determine how fast molten resin is injected into a mold. If either is too high, the resin is forced in so fast that trapped air and gases are not allowed time to be vented. Many of these gases are pushed to the edge of the flow front and become compressed to the point that they auto-ignite, burning the surrounding plastic. The burned areas appear as char marks on the molded part.

Solution: Reducing the injection speed or pressure will allow enough time for the gases or trapped air to escape through normal vent paths.

Excessive Back Pressure

Explanation: While most materials will benefit from some back pressure application, there is a limit to the amount needed for a specific material and product. The whole idea of applying back pressure is to mix the material better, making it denser and more oriented for flow. However, this very act of mixing may introduce air into the melt, which may be too much for the venting system to handle under normal conditions. The excess air may be compressed at the vent locations and auto-ignite, causing burn marks on the part. There is also the possibility that shearing action from too high a back pressure setting will degrade the material in the barrel and cause burning to occur.

Solution: Use minimum back pressure. All materials will benefit from approximately 50-psi back pressure, but some require up to 300 psi. The material supplier is the best source of information regarding proper back pressure settings for a specific material. When adjusting back pressure use increments of 10 psi.

High Screw Rotation Speed

Explanation: The turning of the screw brings fresh material into the heating cylinder and imparts a certain amount of heat to that material through rotary shear friction. The faster the screw turns, the greater the friction and shear heat created. If the speed is too high, the friction will cause the material to overheat and become thermally degraded. This causes small particles of charred material to form and these get pushed into the melt stream during injection. They will show up on the molded part as burned areas.

Solution: Adjust the screw rotation speed. An average speed should be approximately 100 rpm. But, specific materials require specific rotation speeds. Consult the material supplier for the proper speed for a specific resin. When adjusting up or down, do so in increments of 10 rpm.

Improper Compression Ratio of Screw

Explanation: Compression ratio is a value that indicates how much a screw will compress a material while it is being processed in the barrel of the machine. It is calculated by dividing the depth of the feed section (rear) of the screw by the depth of the metering section (front) and is usually less than 2 to 1. If the compression ratio is too great for the specific material being molded, the resin degrades thermally and the burned material flows into the molded part.

Solution: The material supplier can recommend the proper compression ratio for a specific resin. A general-purpose screw can usually be used to provide adequate compression but specific conditions may require a screw that is specially designed for a given material.

Faulty Temperature Controllers and Heater Bands

Explanation: Temperature controllers are sensitive units and should be checked and calibrated often (every 3 months maximum). Heater bands do wear out or burn out and normally there is no way of knowing. If controllers or heater bands are not functioning properly, the other controllers and heater bands must compensate. The result is localized overheating of material in the cylinder. This material can degrade and char and enter the melt stream to become molded into the plastic part.

Solution: Inspect and calibrate temperature controllers at least every 3 months. Inspect and replace damaged heater bands as necessary. In addition, look for broken or crimped wires, poor insulation, rusted areas on the heater bands, and poor electrical connections.

Excessive Barrel Temperatures

Explanation: When barrel temperatures are set too high, the resin will overheat and undergo thermal degradation. This degraded material will break loose, enter the melt stream, and become molded into the finished part.

Solution: Establish proper barrel temperatures and profile. The material supplier will provide accurate barrel temperature requirements. The profile should have the barrel temperatures increase progressively from rear to front.

MOLD

Improper Sprue Bushing-To-Nozzle Sizing

Explanation: If the sprue bushing diameter does not match the nozzle opening (or vice-versa) molecular shearing will occur at their junction and some of the material flowing through that area will degrade. The degraded material will enter the melt stream and be molded into the finished part.

Solution: Using bluing dye or thick paper, press the nozzle against the sprue bushing, and check the impression of the openings of each. They should be close to the same and not be off center. Replace the nozzle tip or the sprue bushing if they do not match. Re-center the heating cylinder to the mold if they are off center.

Improper Venting

Explanation: Air is trapped in a closed mold and incoming molten plastic will compress this air until it auto-ignites. This burns the surrounding plastic and results in charred material in the form of burn marks.

Solution: Vent the mold by grinding thin (0.0005"-0.002") pathways on the shutoff area of the cavity blocks. Vents should take up approximately 30% of the perimeter of the molded part. Vent the runner, too. Any air that is trapped in the runner will be pushed into the part. Blind pockets can be vented using flush core pins or fake ejector pins and grinding a flat down the entire length of the pins.

Undersized Gates

Explanation: Gates are used to determine the flow of the material into the cavity. They are intended to cause a slight restriction and impart shearing heat to the plastic, as well as to control the speed at which the plastic enters. If the gate is too small, the restriction is too great and the material will overheat causing degradation. The degraded material shows up as burned resin in the finished part.

Solution: Size the gate according to the material supplier's recommendations. The gate should be as thin as possible to minimize cycle time, but as thick as necessary to reduce the tendency to degrade material. Gates should be installed in removable inserts so they can easily be altered or replaced.

MATERIAL

Excessive Use of Regrind

Explanation: Regrind melts at a lower temperature than virgin, and a regrind/virgin blend must be heated high enough to melt the virgin, which may degrade the regrind. For this reason, regrind use should be minimized if mixed with virgin material. However, regrind by itself can be used successfully by lowering the melt temperature.

Solution: Use 100% regrind, or, if mixing with virgin, limit the amount of regrind to 15% by weight. It may be necessary to use no regrind at all, especially in some medical and electronic products.

OPERATOR

Inconsistent Process Cycle

Explanation: It is possible that the machine operator is the cause of delayed or inconsistent cycles. This will result in excessive residence time and erratic heating of the material in the injection barrel. If such a condition exists, materials may degrade, resulting in locally burned resin.

Solution: If possible, run the machine on automatic cycle, using the operator only to interrupt the cycle if an emergency occurs. Use a robot if an "operator" is really necessary. And, instruct all employees on the importance of maintaining consistent cycles.

Chapter 16
CLEAR SPOTS

Clear spots can be defined as spotty, transluscent imperfections on the molded part surface and usually are caused by unplasticized particles of resin. They might also be caused by improperly dispersed additives, such as colorant.

Figure 16-1 *Clear spots.*

The following can cause clear spots:

- Low barrel temperature,
- low back pressure,
- slow screw rotation,
- uncontrolled heater bands or thermocouples,
- improper screw compression ratio,
- cracked mold (water leak),
- contamination from mold release,
- contaminated raw material,
- improper regrind use,
- inconsistent process cycle.

MACHINE

Low Barrel Temperature

Explanation: Low injection barrel temperature results in an improper melting and blending of the plastic resin. Some of the unplasticized pellets enter the melt stream and appear in the molded part as small chunks of clear material. The clearer the base resin, the more obvious the particles are, but even dark opaque materials can display these unmelted particles. And, improperly blended additives can have the same general appearance.

Solution: Increasing the barrel temperature will reduce the amount of unplasticized material. It will also help to mix additives with the base resin. Increase the barrel temperature by adding 10 degrees F to each zone until the clear spots disappear. Wait 10 cycles between adjustments for the barrel heats to stabilize. And, watch for signs of degradation as the material temperature rises.

Low Back Pressure

Explanation: While the heater bands and screw friction usually provide all the required plasticizing, it is usually necessary to use a small amount of back pressure to improve the mix. The shearing action of the back pressure setting adds heat but also helps blend additives into the base resin. Low back pressure will result in improper blending of additives and cold spots of material.

Solution: Use a minimum amount of back pressure but make sure it is enough to do the job. Start at 50 psi and increase in increments of 10 psi until proper blending occurs. Do not exceed 300 psi, as that will degrade the base resin.

Slow Screw Rotation

Explanation: Screw rotation helps to impart shear heat into a material and aids in blending the resin and additives. A slow speed will keep the material from achieving proper heat to fully plasticize and some of the unmelted particles will enter the melt stream.

Solution: The screw rotation speed should be set at the material supplier's recommendation. As a rule-of-thumb you can start at 100 rpm and adjust upwards or downwards in 10-rpm increments until the proper speed is achieved. However, excessive screw speed will degrade the material.

Uncontrolled Heater Bands or Thermocouples

Explanation: Improperly sized or damaged heater bands or thermocouples can cause low barrel temperatures. This will keep some of the pellets from melting all the way and these will show up in the molded part as clear spots.

Solution: Check each heat zone to ensure that all heater bands are working properly, are suitably controlled, accurately sized, and are tight against the barrel. A conductive sealant should be used to ensure full contact with the barrel. Be sure to replace bands with the proper size, voltage, and wattage requirements as stated in the machine manual.

Improper Screw Compression Ratio

Explanation: Compression ratio is a value that indicates how much a screw will compress a material while it is being processed in the barrel of the machine. It is calculated by dividing the depth of the feed section (rear) of the screw by the depth of the metering section (front) and is usually less than 2 to 1. If the compression ratio is too low for the specific material being molded, the resin will not melt properly and unmelted particles will be present. These enter the melt stream and appear as clear spots in the molded part.

Solution: The material supplier can recommend the proper compression ratio for a specific resin. A general-purpose screw can usually be used to provide adequate compression but specific conditions may require a screw that is specially designed for a given material.

MOLD

Cracked Mold (Water Leak)

Explanation: A mold that is cracked due to weak waterlines may cause water droplets to enter the cavity. These will be trapped by incoming molten resin and will show up as clear spots in the molded part.

Solution: Waterlines that are leaking may be welded if the cracking condition is not severe. Tubing can be threaded through the cracked waterline and the water can flow through the tubing much like air within a tire inner tube. Or, a reverse water system can be utilized that sucks water through the mold instead of pressurizing it through the mold.

Contamination from Mold Release

Explanation: Use of mold release may cause a residue pocket that gets picked up and trapped in the melt stream as the material flows through the mold. These trapped pockets will not allow material to bond together and those areas will appear as clear spots in the part.

Solution: Eliminate the reason for the use of mold release. If release is needed, use it sparingly and clean the mold cavity surfaces frequently to remove residue.

MATERIAL

Contaminated Raw Material

Explanation: A common cause of clear spots is contaminated raw material. Often, the raw material containers are left uncovered and pellets from other incompatible resins may get deposited in the container. Because they melt at different temperatures the incompatible pellets may never get hot enough to melt and they will be dragged into the cavity appearing as clear spots.

Solution: This type of contamination can be minimized by dealing with high quality, reputable suppliers and by using good housekeeping practices. Properly trained material handlers will also help reduce contamination.

Improper Regrind Use

Explanation: Although regrind melts at a lower temperature than virgin material, if the regrind pellets are too big or coarse, they may not absorb heat as fast as the virgin pellets and may not fully melt in the barrel. These unmelted particles will form clear spots in the part under the right conditions.

Solution: Use a smaller screen in the granulator. This will create smaller, more uniform regrind pellets that will melt fast enough in the barrel. Also, limit the amount of regrind being used to 15%. At that level, or lower, the regrind tends to get dissipated enough to keep the clear spots from showing.

OPERATOR

Inconsistent Process Cycle

Explanation: It is possible that the machine operator is the cause of delayed or inconsistent cycles. This will result in erratic heating of the material in the injection barrel. If such a condition exists, some pockets of material may not melt all the way and unmelted particles will cause the clear spots.

Solution: If possible, run the machine on automatic cycle, using the operator only to interrupt the cycle if an emergency occurs. Use a robot if an "operator" is really necessary. And, instruct all employees on the importance of maintaining consistent cycles.

Chapter 17
CLOUDY APPEARANCE

Cloudy appearance can be defined as an imperfection resembling a localized cloud formation or dull area. It is most apparent throughout a transparent part but can also be evident on the surface of an opaque part.

Figure 17-1 *Cloudy appearance.*

The following can cause a cloudy appearance:
- Low barrel temperature,
- low back pressure,
- low screw rotation,
- excessive wear between barrel and screw,
- improper venting,
- uneven packing,
- dull finish on mold steel,
- poor mold temperature control,
- excessive moisture,
- inconsistent process cycle.

MACHINE

Low Barrel Temperature

Explanation: Low injection barrel temperature results in plastic particles that are not fully melted and blended with the main melt. The cloudy appearance forms when groups of these particles are located together.

Solution: Increasing the barrel temperature reduces the likelihood of unmelted particles. Make sure the profile is correct by heating the material higher as it travels from rear to front in the barrel.

Low Back Pressure

Explanation: The use of back pressure improves the mix by adding heat and creating a more thorough blending action to the melt through shearing action. If no back pressure is used, there is a risk that not all plastic pellets will be fully melted and mixed. This may result in a cloudy appearance in areas where these particles have grouped.

Solution: Use a minimum back pressure setting of 50 psi. Then, increase in 10-psi increments until the cloudy appearance disappears. Do not exceed 300 psi, and keep watch on the material as the back pressure increases, to avoid degradation.

Low Screw Rotation

Explanation: Screw rotation helps to impart shear heat into a material and aids in blending the resin and additives. A slow speed will keep the material from achieving proper heat to fully plasticize and some of the unmelted particles will enter the melt stream. The cloudy appearance is the result of these unmelted particles.

Solution: The screw rotation speed should be set at the material supplier's recommendation. As a rule-of-thumb, you can start at 100 rpm and adjust upwards or downwards in 10-rpm increments until the proper speed is achieved. However, excessive screw speed will degrade the material.

Excessive Wear Between Barrel and Screw

Explanation: The gap between the inside diameter of the injection barrel and the overall diameter of the screw within the barrel is critical. As a general rule, any gap greater than 0.005" is excessive and will cause a variety of molding problems. The greatest result of a large gap is unplasticized material. This is caused by an inability to create back pressure, and a slipping of the pellets across the screw flights during travel through the barrel. Of course, if the pellets are not properly melted they can cause a cloudy appearance on the part.

Solution: Screws and barrels should be checked periodically (at least every 6 months) and replaced or repaired when worn or damaged. The greatest wear will be at the front of the barrel or screw because that is where the material is discharged. Barrels can be repaired by fitting with a liner and screws can be replated.

MOLD

Improper Venting

Explanation: Air is trapped in a closed mold and incoming molten plastic will push this air towards the edges of the cavity. If the air cannot get out of the mold, it will be compressed but will interfere with the ability of the plastic to push against the steel of the cavity. The plastic will take on a dull, cloudy appearance in those areas because it cannot replicate the mold finish.

Solution: Vent the mold by grinding thin (0.0005"-0.002") pathways on the shutoff area of the cavity blocks. Vents should take up approximately 30% of the perimeter of the molded part. Vent the runner, too. Any air that is trapped in the runner will be pushed into the part.

Uneven Packing

Explanation: Improperly gated parts may result in uneven packing of the plastic in localized areas of the cavity. This would be caused, for example, if the part were gated such that the material flows from a thin section to a thick section. The plastic in the thin section will solidify before enough pressure can be applied to the plastic in the thick section. The molecules in the thick section will not be packed against the steel and will appear cloudy.

Solution: Follow proper guidelines for gating, as recommended by the material supplier. Make sure parts are gated to flow the plastic from thick sections to thin sections. And, make sure runners are properly sized. The diameter of the runner should be larger at the sprue than at the cavity if the runner is greater than three inches in length.

Dull Finish on Mold Steel

Explanation: When constructed, the surface of the mold is first machined, then stoned by hand to remove the machining marks, and polished by hand to remove the stoning marks. The final polishing phase can be extensive or brief depending on how much luster is required on the finished product. If polishing is not consistent, there may be areas that are not as highly polished as others. These duller areas will create a dull, cloudy appearance on the molded part.

Solution: Inspect the mold for consistent polishing. If dull areas exist, they can be selectively polished to match other areas. If the mold was plated, make sure the plating is uniform. Also, clean any area where staining may be present. This can come from out-gassing of the plastic but can be easily removed.

Poor Mold Temperature Control

Explanation: As a rule-of-thumb, a hot mold produces a shiny part and a cold mold will produce a dull part. If the entire mold is cold, the entire part is dull. But, if there is local reduced temperature due to inconsistent water flow through the mold, that area will produce a localized cloudy finish on the molded product.

Solution: Make sure the waterlines produce turbulent flow. Simply feeling the "in" line and the "out" line can check this. There should be no more than 10 degrees F difference between the two lines. If there is more than that 10 degree F difference, it means there is inconsistent flow through the mold and hot and cold pockets will be found. The cold pockets will produce a cloudy finish. Turbulence can be created by increasing the flow of water (measured as gallons per minute) into the mold. Finally, make sure the waterlines are hooked up properly. There may not be any water going to certain sections of the mold. Then, the water in the rest of the mold must be set to a colder value to compensate and will cause cloudy parts.

MATERIAL

Excessive Moisture

Explanation: Excessive moisture is a frequent cause of cloudy appearances. Moisture turns to steam when heated in the injection unit, and this steam explodes throughout the plastic, causing voided areas between molecules. The voided areas are not packed together and appear cloudy on the part.

Solution: Although it is commonly understood that non-hygroscopic materials do not require drying, do not take chances. Dry all materials. It may be that fillers used in the material are hygroscopic and they will absorb moisture. Every plastic material requires specific drying conditions, and each material should be dried according to the material supplier's recommendations. The desired moisture content is between 1/10th of 1 percent and 1/20th of 1 percent by weight. This means the dry air being used to take moisture from the material should have a dew point of -20 to -40 degrees F.

OPERATOR

Inconsistent Process Cycle

Explanation: It is possible that the machine operator is the cause of delayed or inconsistent cycles. This will result in erratic heating of the material in the injection barrel. If such a condition exists, some pellets will not be properly plasticized and will enter the melt stream to create cloudy areas in the molded part.

Solution: If possible, run the machine on automatic cycle, using the operator only to interrupt the cycle if an emergency occurs. Use a robot if an "operator" is really necessary. And, instruct all employees on the importance of maintaining consistent cycles.

Chapter 18

CONTAMINATION

Contamination can be defined as an imperfection in a molded part caused by the presence of a foreign object or material that is not part of the original molding compound.

Figure 18-1 *Contamination.*

The following can cause contamination:

- Oil leaks and grease drips,
- uncovered hopper,
- improperly cleaned hopper,
- excessive lubrication,
- contaminated raw material,
- improper regrind,
- excessive moisture,
- poor housekeeping.

MACHINE

Oil Leaks and Grease Drips

Explanation: It is common for most molders to allow small oil leaks to become big ones before they consider repairing them. This oil can find its way into unusual locations, such as a container of raw plastic being prepared to go into the machine hopper. Also, when equipment is greased, it is usually overdone and grease drips can fall into material containers. These contaminants are not compatible with the base resin so they are very evident in the molded part.

Solution: Fix oil leaks as quickly as possible. Clean up grease drips as they occur and do not use the same cleanup rags to wipe out the hopper between material changes.

Uncovered Hopper

Explanation: When the molding machine is first built, it is supplied with a cover for the hopper. It is there to keep contaminants from getting to the molding compound. If that cover is not present during production runs, the material can be contaminated with such things as dust from the ceiling, drops of condensation from overhead waterlines, and even bird droppings in large facilities.

Solution: Use the hopper lid. Do not improvise with flattened cardboard boxes, as the paper particles will cause contamination. If the original lid is lost buy a new one. They are designed to do a specific job well.

Improperly Cleaned Hopper

Explanation: When a material change is required, the hopper must be cleaned out. It is not good enough to simply remove the material present. The sides of the hopper must be wiped to remove any material dust or "fines" that stick to the sides due to static charges. If not removed, these fines will get picked up by the new material and, because of incompatibility, will appear as contamination in the molded parts.

Solution: Clean the hopper thoroughly between material changeovers. This may require wiping the inside with a cloth slightly dampened with denatured alcohol to remove all traces of fines.

MOLD

Excessive Lubrication

Explanation: Molds with cams, slides, lifters, and other mechanical actions need periodic lubrication. Sometimes there is a tendency to overdo this and the lubricant may find its way into the cavity of the mold. This is especially true for ejector pin lubricants. The oil-based lubricant is not compatible with the base resin and is evident as contamination on the molded parts. In addition, excessive mold release acts as a contaminant and appears on the molded part as blotches, dark spots, smears, and streaks.

Solution: Optimize the use of lubricants and minimize the use of mold release sprays. Clean up any excess lubricants and use only the amount needed for a specific application. A little lubricant goes a very long way. Investigate the reason for using mold release. Usually it is a temporary approach to a more severe problem, and the problem should be solved to eliminate the need for the release agent.

MATERIAL

Contaminated Raw Material

Explanation: The most common cause of contamination in molded parts is molding compound contamination. Such contamination is usually the result of dirty regrind, improperly cleaned hoppers or granulators, open or uncovered material containers, and poor quality virgin material supplied by the manufacturer.

Solution: This type of contamination can be minimized by dealing with high quality, reputable suppliers and by using good housekeeping practices. Properly trained material handlers will also help reduce contamination.

Improper Regrind

Explanation: Regrind can be defined as any virgin material that has seen at least one heat excursion through the molding machine. It can be in the form of molded parts or runner systems.

Solution: Grinding the plastic in a granulator that is specially designed for the purpose creates the regrind. The material is ground until it falls through a screen with specifically sized openings. The larger the openings, the larger the regrind pellet. If the granulator is not cleaned out between changes (including the screen), pellets of incompatible materials may be mixed with the next material and this causes contamination. Granulators must be completely taken apart to perform a thorough cleaning between uses. Do not assume that the granulator will always be used only for a certain material. Plans have a way of changing. As soon as a run is completed, the granulator should be removed from service and cleaned. Use of compressed air, combined with wiping with clean rags will suffice, as long as every component is cleaned, including rotor, blades, container, screen, sidewalls, and feed throat.

Excessive Moisture

Explanation: Excessive moisture should be considered a contaminant. It doesn't belong in the molding compound. Moisture turns to steam when heated in the injection unit, and this steam interferes with molecular bonding. This causes splay, which is a visual defect, but also creates a weak part due to brittleness.

Solution: Although it is commonly understood that non-hygroscopic material does not require drying, do not take chances. Dry all materials. It may be that fillers used in the material are hygroscopic and they will absorb moisture. Every plastic material requires specific drying conditions. And each material should be dried according to the material suppliers recommendations. The desired moisture content is between 1/10th of 1 percent and 1/20th of 1 percent by weight. This means the dry air being used to take moisture from the material should have a dew point of -20 to -40 degrees F.

OPERATOR

Poor Housekeeping

Explanation: The machine operator may be the source of contamination in a variety of ways. First, if the operator is allowed to have food or drink at the work station, these may accidentally get spilled into containers holding fresh material ready to go into the hopper. Second, the operator may have been instructed to keep the area clean and sweeping dust into the air may result in contaminating raw material or freshly molded parts. Third, a lack of concern or outright sabotage could be the incentive to purposely add contaminants to the molding compound.

Solution: If possible, run the machine on automatic cycle, using the operator only to interrupt the cycle if an emergency occurs. Use a robot if an "operator" is really necessary. And, instruct all employees on the importance of maintaining contaminant-free areas.

Chapter 19

CRACKING/CRAZING

Cracking and/or crazing can be defined as a fracture or surface breakage in the material of a molded part, usually found in weld line areas, but also on the surface in general.

Figure 19-1 *Cracking and crazing.*

The following can cause cracking and crazing:

- Molded-in stresses,
- cooling time too short,
- undercuts or parting line burrs,
- insufficient draft allowance,
- use of mold release,
- improper ejector design,
- degraded material,
- excessive moisture,
- poor work habits.

MACHINE

Molded-In Stresses

Explanation: Although stress cannot be eliminated, excessive stress can be molded into the parts by using too high an injection pressure, too high a holding pressure, or too fast a filling rate. The molten material is forced into the mold and held there under great pressure until it partially solidifies. When the mold opens and the part is ejected, it is still cooling but is no longer constrained by the mold. Some of the molded-in stresses are allowed to release and a "splitting" of the plastic occurs, usually in the weakest area.

Solution: Reduce packing and fill rates by adjusting until the part is properly filled with minimum stresses. Maintain a holding pressure that is no more than 1/2 the primary injection pressure.

Cooling Time Too Short

Explanation: If the cooling time is too short, the part is ejected before the material has formed a skin solid enough to constrain movement of the remaining plastic material while it cools. The surface will split open and form crazing, or cracks will form throughout the part.

Solution: Increase the cooling time portion of the cycle. This holds the mold closed longer and allows a thicker skin to form on the molded part. The skin will be strong enough to keep crazing or cracks from forming.

MOLD

Undercuts or Parting Line Burrs

Explanation: For the molded part to eject there must be no restrictions to a straight push out of the cavity. An undercut, reverse draft, or burr will cause such a restriction. This will try to keep the part in the mold while the ejection system tries to push it out of the mold. The conflict that arises will cause the part to fracture or crack.

Solution: Inspect the sidewalls and edges of the cavity. Make sure there is adequate draft (see the next paragraph) and that there are no burrs or other undercut conditions. If there are they must be removed by stoning or machining.

Insufficient Draft Allowance

Explanation: A draft angle is simply a tapered side wall that is used to allow easy removal of the molded part from the cavity of the mold that forms it.

Without this taper the vacuum that is created (when plastic displaces air) in the cavity cannot be overcome and the part will simply crack as the ejection system tries to push it out of the mold.

Solution: As a general rule-of-thumb, a draft angle should be at least 1 degree per side to facilitate easy ejection. This does result in a dimensional change in the part and must be considered in the mold design phase. To minimize future problems, the product designer should be made aware of this requirement.

Use Of Mold Release

Explanation: Mold release will interfere with the molecular bonding of the plastic. Material enters a cavity in layers and these layers must be allowed to bond together. Mold release interferes with that bonding and will cause crazing to occur on the surface of the part.

Solution: The remedy is to keep the mold as clean as possible and make every effort to eliminate the use of external mold releases.

Improper Ejector Design

Explanation: If ejector pins are too small, or located on thin flat sections of the part, the plastic will be distorted during ejection and cracks will form due to the amount of stress being imparted. Also, if the ejection speed is too great, the plastic will not have time to conform to the normal stress being applied and cracks will form from distortion.

Solution: Any evidence of cracking in ejector pin areas indicate the pins are too small, or the ejection speed was too great, or there was too much injection pressure used to fill the mold. Injection pressure was covered earlier. Ejector pins should be resized, or relocated. They need to be located such that they are close to side walls or under bosses or other strong areas of the part that can absorb the ejection stresses being produced. They need to be as large as possible in diameter to distribute the ejector forces over a large area.

MATERIAL

Degraded Material

Explanation: One common cause of cracking is the use of material that has become degraded. This can be the result of overheating in the barrel, but a more common cause is the use of bad regrind. Regrind that has been used over and over can easily become degraded to the continued exposures to elevated temperature. It melts at lower temperatures than virgin so the regrind can degrade in the barrel, which must be heated high enough to melt the virgin thereby degrading the regrind. Degraded material is weak and does not have a good molecular bonding of molecules. This results in cracking when the part is exposed to any stress, such as that of the ejection system.

Solution: Use only high grade regrind and use it only once. Mix regrind with virgin at a level of approximately 15% regrind by weight to minimize the tendency to degrade. If this is still a problem, eliminate the use of regrind altogether.

Excessive Moisture

Explanation: Excessive moisture causes cracking or crazing because the water droplets actually turn to steam when heated in the injection unit, and these steam pockets erupt causing voided areas between molecules. This causes those areas to be extremely weak and brittle. The voided areas easily break apart once the mold opens and relieves constraint conditions.

Solution: Although it is commonly understood that non-hygroscopic material do not require drying, do not take chances. Dry all materials. It may be that fillers used in the material are hygroscopic and they will absorb moisture. Every plastic material requires specific drying conditions. And each material should be dried according to the material suppliers recommendations. The desired moisture content is between 1/10th of 1 percent and 1/20th of 1 percent by weight. This means the dry air being used to take moisture from the material should have a dew point of -20 to -40 degrees F.

OPERATOR

Poor Habits

Explanation: Machine operators who have been told to use mold release sprays sparingly, will eventually overuse the spray. The thought seems to be that if a little bit works, a lot will work better. Excessive mold release will interfere with molecular bonding of the plastic and cause weak areas that break apart easily.

Solution: If possible, run the machine on automatic cycle, using the operator only to interrupt the cycle if an emergency occurs. Use a robot if an "operator" is really necessary. And, instruct all employees on the importance of maintaining consistent cycles.

Chapter 20

DELAMINATION

Delamination can be defined as a separation of the layers within a molded part. It can also appear as fish-scaling.

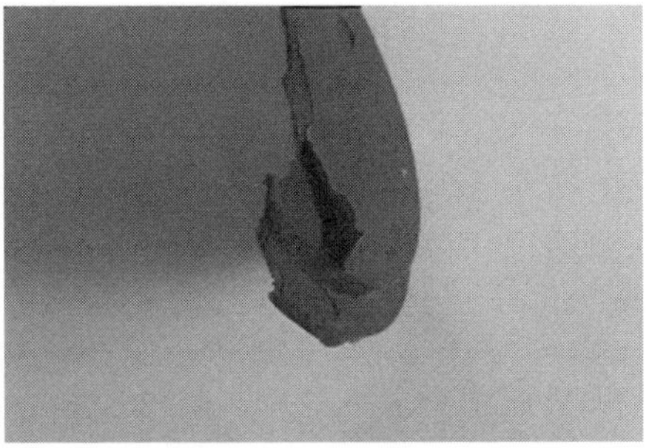

Figure 20-1 *Delamination.*

The following can cause delamination:

- Injection speed too slow,
- inadequate cushion,
- injection "hold" time too short,
- low barrel temperature,
- low mold temperature,
- sharp corners in gates and runners,
- excessive mold release,
- contaminated raw material,
- foreign materials or improper additives,
- excessive moisture,
- poor habits.

MACHINE

Injection Speed Too Slow

Explanation: The injection speed determines how fast a molten material is injected into the mold. The material enters the cavity in layers. If the speed is too slow, the layers will solidify before they are packed and fused together. The weak bonding of these layers results in separation as the part cools and shrinks.

Solution: Increase the injection fill speed in small increments (two percent of total) until the delamination tendency disappears. If burning or flashing occurs, the injection speed is too high and delamination is being caused by other factors.

Inadequate Cushion

Explanation: A cushion, or pad, of material is needed at the end of the injection stroke so pressure can be held at all times on the material in the mold as it solidifies. This cushion is created by adding a little more material than is actually needed to fill the mold. Without that cushion, the screw will bottom out at the end of injection and will be unable to hold pressure against the material in the cavity. That material will not be packed out and the layers that were formed during injection will not fuse together. As the part cools and shrinks outside of the mold (after ejection), the layers will pull apart causing delamination.

Solution: Establish a cushion of at least 1/8" and not more than 1/4". Too little a cushion causes under-packing while too great a cushion results in uncontrolled packing of the sprue area.

Injection "Hold" Time Too Short

Explanation: Holding pressure is applied at the end of the injection stroke. The holding pressure is used for maintaining pressure against the molten material to pack it together while the skin solidifies. If the holding pressure is stopped too early, the still-molten material will leak back out of the cavity and pressure will be lost. Without the pressure, the layers of plastic cannot be fused together and they will pull apart as the molded part cools and shrinks.

Solution: Maintain the holding pressure until the gate freezes. Once that happens, the holding pressure will no longer have any effect on the material in the cavity.

Low Barrel Temperature

Explanation: If the material is not heated to high enough temperature in the barrel, the molecules will not bond together properly. When this material is injected into the cavity, it simply pulls apart and delaminates during the cooling and shrinking phase because it is not held together properly to begin with.

Solution: Each material has a range in which it can be molded and the barrel temperature should be adjusted to accommodate the specific material being used and the product and mold design being incorporated. The material supplier is the best source of information for this data. Be sure to use the proper profile and heat the material progressively from the rear to the front of the barrel.

MOLD

Low Mold Temperature

Explanation: If the mold temperature is too low, it will cause the first layers of molten plastic to solidify too soon. The next layers will not fuse properly to the first layers and delamination will occur when the part cools and shrinks.

Solution: Each material has an ideal mold temperature that should be used for proper molding. The material supplier is the best source for this information, but most material should run in molds that are between 120 - 180 degrees F.

Sharp Corners In Gates and Runners

Explanation: Sharp corners, or 90 degree bends that are positioned along the flow path of runner system and gate, will shear the plastic as it goes by and tear the molecules apart. Breaking this bond results in the individual layers cooling at different rates and not bonding back together again in a proper manner. Delamination occurs when the part cools and shrinks.

Solution: It is always a good idea to utilize a radius at every turn in a runner system and at each gate entrance to the cavity. This will help to minimize shear and reduce the chance of resin separation during injection.

Excessive Mold Release

Explanation: Excessive use of mold release will interfere with the normal bonding process that must occur between layers of plastic as they enter the cavity of the mold. If the layers are not properly bonded and fused they will pull apart as the molded product cools and shrinks after ejection from the mold.

Solution: The remedy is to keep the mold as clean as possible and make every effort to eliminate the use of external mold releases.

MATERIAL

Contaminated Raw Material

Explanation: Some common causes of delamination are dirty regrind, improperly cleaned hoppers or granulators, open or uncovered material containers, and poor quality virgin material supplied by the manufacturer. Contaminants are considered non-compatible with the base resin and will interfere with bonding of layers. The non-fused layers pull apart after the part is ejected from the mold.

Solution: This type of contamination can be minimized by dealing with high quality, reputable suppliers and by using good housekeeping practices. Properly trained material handlers will also help reduce contamination.

Foreign Materials or Improper Additives

Explanation: If a pigment is used to color the resin, it is possible that a non-compatible soap is used in its manufacture. If a concentrate is used for coloring, the base resin may not be compatible with the molding compound being used. Accidental mixing of two different grades of the same material may have occurred and they are not necessarily compatible. Even different flow grades of the same material may not be compatible enough to bond properly. These situations all lead to poor bonding of the plastic layers being formed during the injection process. Improper bonding results in the layers pulling apart during the cooling phase.

Solution: Use only compatible additives for all base resins. Use reputable, proven compounders when specifying non-standard formulations.

Excessive Moisture

Explanation: Excessive moisture is one cause of delamination. It occurs because the water droplets actually turn to steam when heated in the injection unit, and this steam explodes throughout the plastic, interfering with molecular bonding, causing voided areas between molecules. This causes those areas to be extremely weak and brittle. The voided areas easily pull apart and delaminate when the part cools and shrinks.

Solution: Although it is commonly understood that non-hygroscopic material does not require drying, do not take chances. Dry all materials. It may be that fillers used in the material ARE hygroscopic and they will absorb moisture. Every plastic material requires specific drying conditions. And each material should be dried according to the material suppliers recommendations. The desired moisture content is between 1/10th of 1 percent and 1/20th of 1 percent by weight. This means the dry air being used to take moisture from the material should have a dew point of -20 to -40 degrees F.

OPERATOR

Poor Habits

Explanation: Machine operators, who have been told to use mold release sprays sparingly, will eventually overuse the spray. The thought seems to be that if a little bit works, a lot will work better. Excessive mold release will interfere with molecular bonding of the plastic and cause weak layers that pull apart easily.

Solution: If possible, run the machine on automatic cycle, using the operator only to interrupt the cycle if an emergency occurs. Use a robot if an "operator" is really necessary. And, instruct all employees on the importance of maintaining consistent cycles.

Chapter 21
DISCOLORATION

Discoloration can be defined as a change in the original color of a plastic material usually caused by overheating, mechanical shear, contamination, or chemical attack.

Figure 21 *Discoloration.*

The following can cause discoloration:

- Excessive shot size ratio,
- excessive residence time,
- high barrel temperature,
- nozzle temperature too high,
- excessive cycle time,
- improper screw design,
- improper mold temperature,
- inefficient cooling,
- contaminated raw material,
- inconsistent process cycle.

MACHINE

Excessive Shot Size Ratio

Explanation: Ideally, a shot size should equal 50% of the capacity of the barrel. That results in processing one cycle while preparing for the next cycle. However, this is a general statement because, depending on the material, the ratio can be as small as 20% for non-heat-sensitive materials such as polypropylene, and up to 80% for heat-sensitive material such as PVC. As the ratio drops, the residence time of the material in the barrel increases, as does the risk of thermal degradation. Degraded material will discolor due to the molecular breakdown. The material will start showing dark yellowing that progresses towards black as the degradation worsens.

Solution: Strive for a 50% shot-to-barrel ratio. This is ideal but can go as low as 20% if the material is not too heat sensitive (like polypropylene) and up to 80% if the material is extremely heat sensitive (like PVC). It is not a good idea to empty the barrel for every shot as more time will be required to bring the next mass of material up to proper heat and degradation may occur.

Excessive Residence Time

Explanation: The longer a material resides in the barrel before being injected, the more heat the material absorbs. The barrel temperatures are set based on the cycle time of the machine. If that cycle time increases or is interrupted, the material residence time increases. That increase may degrade the material and cause discoloration due to the molecular breakdown. In addition, if the mold is in a press that is too large, the residence time for the material will be too great and degradation will occur.

Solution: Size the mold to run in a press that supplies an injection shot size of 50% of the barrel if possible. Optimize the overall cycle of the machine, and eliminate any interruptions to that cycle to minimize the amount of time the material is in the heated barrel.

High Barrel Temperature

Explanation: When barrel temperatures are too high, the material will overheat and undergo thermal degradation. Due to the tendency for the material to carbonize, the more it degrades, the darker it becomes. This results in discoloration in varying degrees from slight to extreme.

Solution: Reduce the barrel temperature to the range recommended by the material supplier. Make sure the profile is such that the material heats progressively from the rear to the front of the barrel.

Nozzle Temperature Too High

Explanation: As material is transported through the heating barrel, it is gradually brought up to the ideal processing temperature. The material absorbs heat from the heating bands and frictional heat, which is created by the shearing action of the rotating screw within the barrel. The last heating zone that the material is exposed to is the nozzle. By the time the material reaches the nozzle, it should already be at ideal molding temperature and any further heat introduced may cause the material to begin to degrade. The degraded material darkens as it becomes more degraded and discolors the molded part.

Solution: Reduce the nozzle temperature to be the same as, or 10 degrees F hotter than, the front zone of the barrel. The extra 10 degrees is used to make up for any heat loss occurring between the nozzle and the sprue bushing against which it seats.

Excessive Cycle Time

Explanation: A cycle time that is too long will increase the residence time for the material in the heating cylinder. The longer residence time may cause the material to begin to degrade and discoloration will occur as the material begins to carbonize.

Solution: Optimize the cycle time to accommodate the material being used and the wall thickness of the part being molded. If longer cycles are actually needed, reduce the barrel temperature to minimize degradation.

Improper Screw Design

Explanation: Injection molding machines are furnished originally with a "general purpose" screw. This screw is designed to perform adequately with most materials. However, each material has a specific optimized screw design available. If the screw is not properly designed for a heat sensitive material, the resin may be degraded by the screw action. Degraded material causes discoloration in the molded part.

Solution: The compression ratio of the screw should be right for the material being molded. The material supplier can provide the ideal compression ratio and the screw manufacturer can provide the compression ratio of the screw being used. If the ratio is not the same, you may have to purchase a screw with the right compression ratio, especially for heat sensitive materials.

MOLD

Improper Mold Temperature

Explanation: A hot mold will cause a molded part to be darker than if it had been molded in a cold mold. This is because the hotter mold allows the material molecules to pack tighter before they solidify and the part is more dense. The colder mold causes the material to solidify before the molecules are packed tightly so the part is less dense.

Solution: Adjust the mold temperature to that recommended by the material supplier. If the parts are too light, heat the mold more. If the parts are too dark, reduce the mold temperature. Make adjustments in 10 degree increments and allow 10 cycles between adjustments for the machine to stabilize.

Inefficient Cooling

Explanation: Anything that impairs the cooling process within a mold will have a direct effect on the color of the molded parts. A blocked or kinked waterline will cause sections of the mold to run too hot and this will cause darker areas to be molded on the part. If there are not enough waterlines the mold will be too hot in general and the whole part will be molded darker because it is denser. If there is too much cooling in one area, that will produce a lighter area of color on the molded part.

Solution: The cooling lines should be designed from the beginning to be efficient and properly located. This is the job of the mold designer. If the mold was not designed right it can be a major problem trying to accommodate for that in the molding process. One thing that can be done is to ensure each mold half has its own temperature control system and that a single unit does not control both halves.

MATERIAL

Contaminated Raw Material

Explanation: If the raw material contains any contamination, such as dirty regrind, dust from storage areas, etc., the molded part will show varying degrees of discoloration based on the location and type of contaminant. Even mixing different grades of the same material or different flow values of the same material can cause differences in color.

Solution: The solution is to keep all materials clean, covered, and stored in proper containers with the contents clearly identified, and make sure grade and melt flow are included.

OPERATOR

Inconsistent Process Cycle

Explanation: It is possible that the machine operator is the cause of delayed or inconsistent cycles. This will result in excessive residence time of the material in the injection barrel. If such a condition exists, heat sensitive materials will degrade, resulting in discoloration.

Solution: If possible, run the machine on automatic cycle, using the operator only to interrupt the cycle if an emergency occurs. Use a robot if an "operator" is really necessary. And, instruct all employees on the importance of maintaining consistent cycles.

Chapter 22
FLASH

Flash can be defined as excess plastic material forced out of the cavity. This usually occurs at the parting line, but can be at any point where two mold surfaces meet.

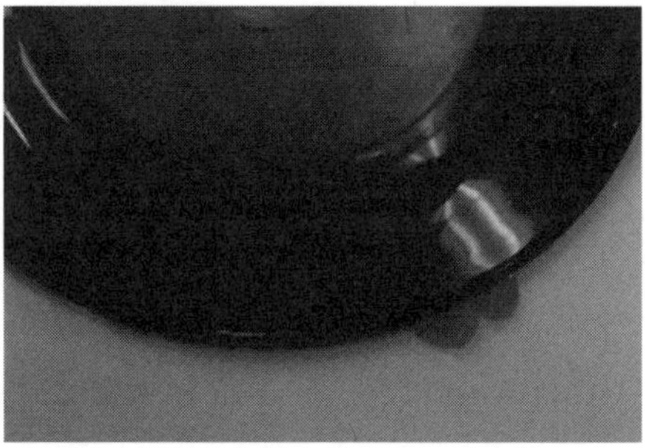

Figure 22-1 *Flash.*

The following can cause flash:
- Excessive residence time in barrel,
- excessive injection pressure,
- high barrel temperatures,
- excessive cycle times,
- inadequate clamp force,
- improper parting line seal,
- improper venting,
- inadequate mold supports,
- sprue bushing too long,
- improper stackup dimensions,
- improper flow rate,
- excessive mold lubricant,
- inconsistent process cycle.

MACHINE

Excessive Residence Time In Barrel

Explanation: Ideally, a shot size should equal 50% of the capacity of the barrel. That results in processing one cycle while preparing for the next cycle. However, this is a general statement because, depending on the material, the ratio can be as small as 20% for non-heat-sensitive materials such as polypropylene and up to 80% for heat-sensitive material such as PVC. As the ratio drops, the time of residence by the material in the barrel increases, and the material gets hotter. It will flow much easier at this point and enter areas where it could not at its normal viscosity. This results in a flashing condition in those areas.

Solution: Strive for a 50% shot-to-barrel ratio. This is ideal but can go as low as 20% if the material is not too heat sensitive (like polypropylene) and up to 80% if the material is extremely heat sensitive (like PVC). It is not a good idea to empty the barrel every shot because more time will be required to bring the next mass of material up to proper heat and degradation may occur.

Excessive Injection Pressure

Explanation: If too much injection pressure is used, the clamp unit may not be able to hold the mold closed, especially on a machine with a hydraulic clamp. The mold will blow open and plastic material will leak out around the perimeter of the parting line. This leakage is called *flash*. It can also occur between any mating steel surfaces of the mold such as around ejector pins and slides. Major mold damage may occur as a result of flashing.

Solution: Reducing the injection pressure reduces the tendency for the material to flash. In addition, make sure there is a properly raised shutoff land around the perimeter of the cavity. This will focus the clamp force and allow less clamp tonnage to be used. Without the shutoff land on the mold, the machine may not be able to generate enough clamp force to keep the mold closed under normal injection pressure.

High Barrel Temperatures

Explanation: High barrel temperatures have the same effect as long residence time. The plastic material becomes more fluid than it should and can enter small openings and crevices that it could not normally enter at the right viscosity.

Solution: Reduce the barrel temperature to that recommended by the material supplier. And, remember to keep the profile set so the material is heated from the rear towards the front of the barrel.

Excessive Cycle Times

Explanation: If the total cycle time is too long, there is a good possibility that the material is overheating in the barrel. This will cause the material to become too fluid and creates the potential for flashing in areas where it would not normally do so.

Solution: Reduce the cycle time. Normally, this can come from the cooling portion of the cycle, but make sure the other functions are not excessive. For instance, injection hold time only needs to be long enough for the gate to freeze. After that, the hold pressure has no effect on the material in the cavity. So, the hold time is an area that should be considered for time reduction. Other functions should also be analyzed.

Inadequate Clamp Force

Explanation: In both hydraulic and mechanical clamp machines, the clamp unit position must be set at the beginning of a molding run and readjusted as the run progresses, due to thermal expansion of the mold and machine. If this readjustment requirement is ignored, the clamp unit position may shift to the point of not fully closing the mold against the incoming injection pressure, and flashing will occur at the parting line. Also, it is important to make sure the mold is placed in a machine that is large enough to provide the required clamp force. If the molding surface area is too great for the size of machine being used, the clamp will not provide enough pressure to hold the mold closed.

Solution: Size the mold to run in the proper machine. This is done by calculating the molding surface area (area of the part to be molded) and multiplying it by a factor of from 2 to 6. The higher number is used for stiff material (like polycarbonate) and the lower number for easy-flowing materials. That will give the number of tons needed to keep the mold closed, assuming there is a proper shutoff land on the mold. You must calculate the total area so include all cavities and the runner system.

MOLD

Improper Parting Line Seal

Explanation: The parting line(s) of a mold must be machined to very close tolerances and parallelism to seal properly when the mold is clamped shut. f the parting line is not parallel, or is otherwise improperly machined or designed, molten plastic material will be forced out of the areas that are not closed tightly, and flash will form.

Solution: Check for proper parting line seal. Make sure there is a shutoff land around the perimeter of the part. There should also be pads around the leader pins at the same height as the shutoff land to ensure parallelism when the mold is clamped. Use a dial indicator to check the flatness (or parallelism) of the parting line surfaces. They should be within 0.002" (or less) over the entire parting line surface.

Improper Venting

Explanation: If vents are machined too deeply for a specific material the molten plastic can leak into the vents and become flash on the molded part.

Solution: Vent the mold by grinding thin (0.0005"-0.002") pathways on the shutoff area of the cavity blocks. The viscosity of the plastic being molded determines the depth of the vent. Stiff materials can utilize deeper vents but fluid materials require thinner vents. In either case, the concept is to remove air from the mold as fast as possible with as deep a gate as the material viscosity will allow.

Inadequate Mold Supports

Explanation: Components called *support pillars* are used in the construction of a mold to provide extra compression support behind the cavity retainer plates on the ejector half of the mold. These pillars are used to fill in the vacant areas present in the U-shaped ejector housing and, when properly positioned, keep the mold from collapsing into the ejector housing during the injection phase of the molding cycle. If there are too few pillars or they are not positioned properly, the mold plates will deflect when injection pressure is applied and molten material will flow into the distorted areas, causing flash.

Solution: Ensure that adequate support exists. An example of the importance of support pillars can be seen by the following: If a 12" x 15" mold base is used without any pillars, the maximum amount of projected part area that the mold could produce without plates deflecting would be 14 square inches. If four 1-1/4" diameter support pillars are properly placed in the same mold, the allowed projected area would increase to 56 square inches, an improvement of 400%.

Sprue Bushing Too Long

Explanation: In a standard mold design, the sprue bushing extends through the A half of the mold until it touches the parting line at the "B" half. A runner is then machined across the face of the bushing to allow molten plastic to flow into the mold. If the sprue bushing is too long, it will keep the "B" half from closing tightly against the A half and a gap will form at the parting line. Molten material will leak into this gap causing flash.

Solution: Reduce the length of the sprue bushing. This is easily done by grinding the face back enough to form a small pad of material to ensure the bushing does not touch against the "B" half. The thickness of the pad should be limited to approximately 1/32" so it will not affect the overall cooling time of the cycle. This pad will also act as a cold well.

Improper Stack-Up Dimensions

Explanation: Unless a mold is cut-in-the-solid, there are many plates, blocks, and other components used in its construction. Each half of the mold must be constructed so that these items "stack-up" to specific dimensions. If this is not the case, these items will have gaps that the molten material can flow to and cause flash.

Solution: A new mold should have the dimensions checked and adjusted even before the mold is placed in a press. As molds age, the components are exposed to compression and fatigue and may relax. They need to be adjusted periodically to ensure that the stack-up dimensions are still proper. Proper stack-up results in a preload of approximately 0.003" on the cavity block faces.

MATERIAL

Improper Flow Rate

Explanation: Resin manufacturers supply specific formulations in a range of standard flow rates. Thin-walled products may require an easy flow material while thick-walled products can use a material that has a stiffer consistency. It is better to use the stiffest flow possible because it improves physical properties of the molded part. However, the stiff material will require higher injection pressures, which may blow the mold open and cause flash at the parting line. If an easy flow material is used, the physical properties will not be as great. In addition, the material will flow into very thin areas and could create flash where the stiffer materials would not.

Solution: Utilize a material that has the stiffest flow possible without causing non-fill. Contact the material supplier for help in deciding which flow rate should be used for a specific application.

Excessive Mold Lubricant

Explanation: If a material is stiff, a lubricant can be added to improve the flow. If this is an external lubricant such as a mold release agent, it is difficult to control the amount of lubricant being used and the material may become more fluid than required. The result could be flashing where the material would not do so without lubricant.

Solution: If it a lubricant must be used, have the material manufacturer (or a compounder) add it directly to the pellets. That will result in more uniform blending and all the material will have the same flow rate.

OPERATOR

Inconsistent Process Cycle

Explanation: It is possible that the machine operator is the cause of delayed or inconsistent cycles. This will result in erratic heating of the material in the injection barrel. If such a condition exists, the material will not be of consistent flow rate and the easier flowing portions may cause flashing.

Solution: If possible, run the machine on automatic cycle, using the operator only to interrupt the cycle if an emergency occurs. Use a robot if an "operator" is really necessary. And, instruct all employees on the importance of maintaining consistent cycles.

Chapter 23
FLOW LINES

Flow lines can be defined as linear grooving, or circular ripples, on the surface of a molded part that indicate the direction of material flow within the cavity of the mold.

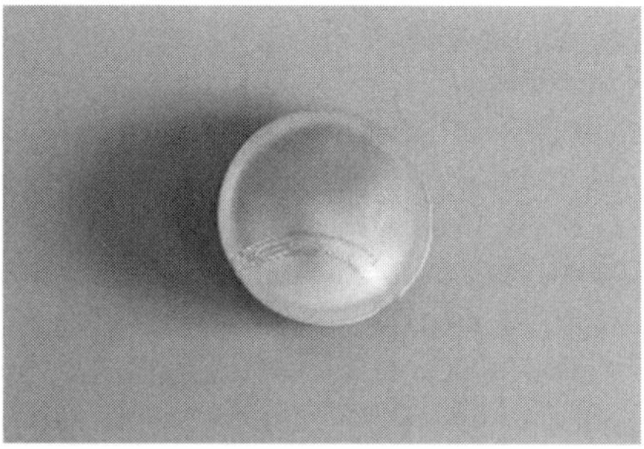

Figure 23 *Flow lines.*

The following can cause flow lines:
- Inadequate injection pressure,
- inadequate residence time,
- low barrel temperatures,
- low nozzle temperature,
- inadequate cycle time,
- low mold temperature,
- improper venting,
- small gates and/or runners,
- improper flow rate,
- inadequate lubricant,
- inconsistent process.

MACHINE

Inadequate Injection Pressure

Explanation: If too little injection pressure is used, the molten plastic will tend to cool down and solidify before the mold is packed out. If no packing is achieved, the flow pattern of the material will be imprinted on the surface of the part because not enough pressure was used to force the plastic against the steel of the mold and squeeze out the flow lines.

Solution: Increasing the injection pressure will force the molten plastic against the mold cavity steel before the plastic solidifies, removing the flow lines and duplicating the cavity finish.

Inadequate Residence Time

Explanation: Residence time is the amount of time that the plastic material spends being exposed to heating conditions in the injection barrel. The required time depends upon how much heat the material must absorb to be processed properly. Inadequate residence time results in underheated material. This will cause the material to be stiff when injected and it will not flow enough to fill the cavity before solidifying. The flow patterns will be imprinted on the surface of the molded part because they were not forced out in time.

Solution: Optimize the residence time by making sure the mold is sized to the proper machine. Also, optimize the cycle time to ensure the material residence time is adequate to properly melt the plastic.

Low Barrel Temperatures

Explanation: Low barrel temperatures have the same effect as short residence time. The plastic material does not become fluid enough to fill the mold before solidifying and flow lines are imprinted on the part surface before they can be forced away.

Solution: Increase the barrel temperature to that recommended by the material supplier. Adjust as needed to eliminate the flow lines. And, remember to keep the profile set so the material is heated from the rear towards the front of the barrel.

Low Nozzle Temperature

Explanation: As material is transported through the heating barrel, it is gradually brought up to the ideal processing temperature by absorbing heat from the heating bands and frictional heat, which is created by the shearing action of the rotating screw within the barrel. In the last heating zone, the material is exposed to is the nozzle. By the time the material gets to the nozzle, it should already be at ideal molding temperature and only a small amount of heat needs to be applied at this point to keep the resin flowing. If the nozzle is not hot enough, however, the material will begin to cool off too quickly as it leaves the barrel and the flow front will not be forced against the cavity steel to squeeze out the flow lines.

Solution: Increase the nozzle temperature. As a rule-of-thumb, the nozzle temperature should be set at 10 degrees F higher than the setting for the front zone of the barrel. This helps compensate for heat loss due to metal-to-metal contact between the nozzle and the sprue bushing, and keeps the material hot enough to pack the mold, eliminating flow lines.

Inadequate Cycle Time

Explanation: If the overall cycle time is too short there is a good possibility that the material in the barrel cannot absorb enough heat before it is injected into the mold. This will cause premature solidification and flow lines may appear because the plastic was not packed enough (before solidifying) to squeeze them out.

Solution: Increase the cycle time. The easiest change to make is to add time to the cooling portion of the cycle. That is when the plastic is absorbing the most heat in the barrel. Increase barrel temperatures 10 degrees F at a time, allowing 10 cycles between changes to re-stabilize the process.

MOLD

Low Mold Temperature

Explanation: Generally, a hot mold will allow a material to stay molten longer than a cold mold and cause the molecules to pack together properly before they solidify. This results in a dense part with no flow lines. If the mold is too cold, the molecules solidify before they are packed out and flow lines may result.

Solution: Increase the mold temperature to the point that the material has proper flow and packs out the mold. Start with the material suppliers recommendations and adjust accordingly. Allow 10 cycles for every 10-degree change for the process to re-stabilize.

Improper Venting

Explanation: If there is not enough venting in the mold, the material will push into unvented areas and not compress against the mold steel because trapped gases are in the way. The material will actually "stutter" as it tries to force the gas out of the way, and will eventually solidify before packing can be achieved. The stutter marks will imprint on the part surface as flow lines.

Solution: Vent the mold by grinding thin (0.0005"-0.002") pathways on the shutoff area of the cavity blocks. The viscosity of the plastic being molded determines the depth of the vent. Stiff materials can utilize deeper vents but fluid materials require thinner vents. In either case, the concept is to remove air from the mold as fast as possible with as deep a gate as the material viscosity will allow. At least 30% of the parting line perimeter should be vented, but additional vents can be selectively placed for any area where flow lines appear.

Small Gates and/or Runners

Explanation: Gates and/or runners that are too small will cause excessive restriction to the flow of the molten plastic. Many plastics will then begin to solidify before they fill the cavity. The result is an unpacked condition of the molecules and flow lines will not have a chance to be pressed out of the product surface.

Solution: Examine the gates and runners to determine if any burrs or other obstructions exist. If possible, perform a computer analysis to determine the proper sizing and location of gates and runners. Ask the material supplier for data concerning gate and runner dimensioning for a specific material and flow rate.

MATERIAL

Improper Flow Rate

Explanation: Resin manufacturers supply specific formulations in a range of standard flow rates. Thin-walled products may require an easy flow material while thick-walled products can use a material that has a stiffer consistency. It is better to use the stiffest flow possible because it improves physical properties of the molded part. However, the stiff material will require higher injection pressures, which may blow the mold open and cause flash at the parting line. If an easy flow material is used, the physical properties will not be as great but, in addition, the material will flow into very thin areas and could create flash where the stiffer materials would not.

Solution: Utilize a material that has the stiffest flow possible without causing non-fill. Contact the material supplier for help in deciding which flow rate should be used for a specific application.

Inadequate Molding Lubricant

Explanation: If a material is too stiff, it may solidify before packing the cavity and flow lines could exist. A lubricant can be added to improve the flow. If this is an external lubricant such as a mold release agent, it is difficult to control the amount of lubricant being used and the material may more fluid than required. The result could be flashing where the material would not do so without lubricant.

Solution: If it is determined that a lubricant must be used, have the material manufacturer (or a compounder) add it directly to the pellets. That will result in more uniform blending and all the material will have the same flow rate.

OPERATOR

Inconsistent Process Cycle

Explanation: It is possible that the machine operator is the cause of delayed or inconsistent cycles. This will result in erratic heating of the material in the injection barrel. If such a condition exists, the colder particles will require higher injection pressures and may not fill the mold before they fully solidify. Flow lines will not be forced out in time.

Solution: If possible, operate the machine on automatic cycle, using the operator only to interrupt the cycle if an emergency occurs. Use a robot if an "operator" is really necessary. And, instruct all employees on the importance of maintaining consistent cycles.

Chapter 24
GLOSS (LOW)

Low gloss can be defined as a dulling of the product finish, usually caused by insufficient force to push the molten plastic against the steel surface of the cavity.

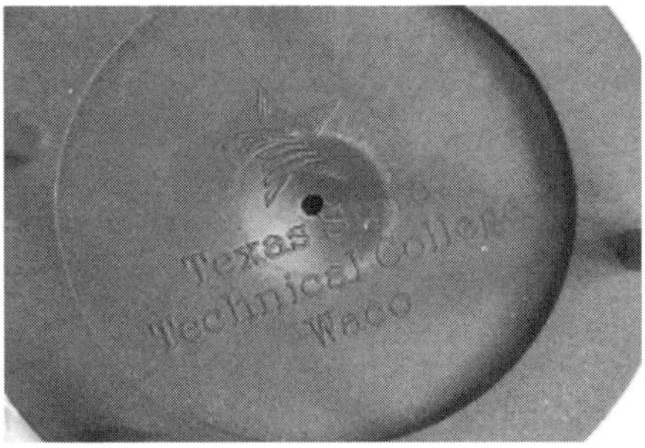

Figure 24-1 *Low gloss.*

The following can cause low gloss:
- Inadequate injection pressure,
- inadequate residence time,
- low barrel temperatures,
- nozzle temperature too low,
- inadequate cycle time,
- low mold temperature,
- small gates and/or runners,
- inadequate venting,
- improper gate location,
- poor polishing of cavity surfaces,
- improper flow rate,
- moisture,
- inconsistent process cycle.

MACHINE

Inadequate Injection Pressure

Explanation: If too little injection pressure is used, the molten plastic material will tend to cool off and solidify before the mold cavity is packed. If no packing is achieved, the material will not be forced against the mold cavity hard enough to replicate the finish. The less dense material will appear dull and have very little gloss.

Solution: Increasing the injection pressure will force the plastic against the steel of the mold cavity and duplicate the gloss of the finish on that steel.

Inadequate Residence Time

Explanation: Residence time is the amount of time that the plastic material spends being exposed to heating conditions in the injection barrel. The required time depends upon how much heat the material must absorb to be processed properly. Inadequate residence time results in underheated material. This will cause the material to be stiff when injected and it will not flow enough to fill the cavity before solidifying. The flow patterns will be imprinted on the surface of the molded part because they were not forced out in time.

Solution: Optimize the residence time by making sure the mold is sized to the proper machine. Also, optimize the cycle time to ensure the material residence time is adequate to properly melt the plastic.

Low Barrel Temperatures

Explanation: Low barrel temperatures have the same effect as short residence time. The plastic material does not become fluid enough to fill the mold before solidifying and the material cannot be forced against the steel of the mold cavity. Therefore, the steel finish will not be replicated on the plastic and a dull part will result.

Solution: Increase the barrel temperature to that recommended by the material supplier. Adjust as needed to eliminate the flow lines. And, remember to keep the profile set so the material is heated from the rear towards the front of the barrel.

Low Nozzle Temperature

Explanation: As material is transported through the heating barrel, it is gradually brought up to the ideal processing temperature by absorbing heat from the heating bands and frictional heat, which is created by the shearing action of the rotating screw within the barrel. In the last heating zone, the material is exposed to is the nozzle. By the time the material gets to the nozzle, it should already be at ideal molding temperature and only a small amount of heat needs to be applied at this point to keep the resin flowing. If the nozzle is not hot enough, however, the material will begin to cool off too quickly as it leaves the barrel and the flow front will not be forced against the cavity steel to squeeze out the flow lines.

Solution: Increase the nozzle temperature. As a rule-of-thumb the nozzle temperature should be set at 10 degrees F higher than the setting for the front zone of the barrel. This helps compensate for heat loss due to metal-to-metal contact between the nozzle and the sprue bushing, and keeps the material hot enough to pack the mold, eliminating low gloss.

Inadequate Cycle Time

Explanation: If the overall cycle time is too short there is a good possibility that the material in the barrel cannot absorb enough heat before it is injected into the mold. This will cause premature solidification and low gloss may appear because the plastic was not packed out enough (before solidifying) to push against the steel cavity of the mold and replicate the steel finish.

Solution: Increase the cycle time. The easiest change to make is to add time to the cooling portion of the cycle. That is when the plastic is absorbing the most heat in the barrel. Increase barrel temperatures 10 degrees F at a time, allowing 10 cycles between changes to re-stabilize the process.

MOLD

Low Mold Temperature

Explanation: Generally, a hot mold will allow a material to stay molten longer than a cold mold and cause the molecules to pack against the mold steel before they solidify. This results in a dense part with high gloss. If the mold is too cold, the molecules solidify before they are packed out and low gloss may result.

Solution: Increase the mold temperature to the point at which the material has the proper flow and packs out the mold. Start with the material suppliers recommendations and adjust accordingly. Allow 10 cycles for every 10-degree change for the process to re-stabilize.

Small Gates and/or Runners

Explanation: Gates and/or runners that are too small will cause excessive restriction to the flow of the molten plastic. Many plastics will then begin to solidify before they fill the cavity. The result is an unpacked condition of the molecules and the material will not have a chance to be pressed against the cavity steel. Low gloss will result.

Solution: Examine the gates and runners to determine if any burrs or other obstructions exist. If possible, perform a computer analysis to determine the proper sizing and location of gates and runners. Ask the material supplier for data concerning gate and runner dimensioning for a specific material and flow rate.

Inadequate Venting

Explanation: If there is not enough venting in the mold, the material will push into the areas that are not vented and will not compress against the mold steel because trapped gases are in the way. The material will actually solidify before packing can be achieved. This early solidification will result in a dull surface finish.

Solution: Vent the mold by grinding thin (0.0005"-0.002") pathways on the shutoff area of the cavity blocks. The viscosity of the plastic being molded determines the depth of the vent. Stiff materials can utilize deeper vents but fluid materials require thinner vents. In either case, the concept is to remove air from the mold as fast as possible with as deep a gate as the material viscosity will allow. At least 30% of the parting line perimeter should be vented, but additional vents can be selectively placed for any area where localized low gloss occurs.

Improper Gate Location

Explanation: With some plastics, if material is injected directly across a flat cavity surface, it tends to slow down quickly as a result of frictional drag and cools off before the cavity is properly filled and packed. When this happens, the molded part surface has low gloss because the material has not been forced against the cavity steel before solidification.

Solution: Relocate or redesign the gate so that the molten plastic is directed against a metal obstruction instead of across a flat surface. This will cause the material to disperse and continue to flow instead of slowing down.

Poor Polishing of Cavity Surfaces

Explanation: A properly molded product will duplicate the finish that is present on the molding surfaces of the mold in which it was formed. If that finish was not properly prepared (normally by hand polishing), the molded part will not have the high gloss that is normally desired.

Solution: Prepare the molding surface finish to the requirements of the molded product. There are industry standards available that describe the degree of gloss required for specific finishes. These should be utilized to ensure consistency of finish on all molded products, and the finish should be specified on the product drawing as well as on the mold design.

MATERIAL

Improper Flow Rate

Explanation: Resin manufacturers supply specific formulations in a range of standard flow rates. Thin-walled products may require an easy flow material while thick-walled products can use a material that has a stiffer consistency. It is better to use the stiffest flow possible because it improves physical properties of the molded part. However, the stiff material will require higher injection pressures, which may blow the mold open and cause loss of pressure, which results in low surface gloss. If an easy flow material is used, the physical properties will not be as great. In addition, the material will maintain pressure and create proper gloss.

Solution: Utilize a material that has the stiffest flow possible without causing non-fill. Contact the material supplier for help in deciding which flow rate should be used for a specific application.

Moisture

Explanation: If moisture is present in the material it will turn to steam during the transition through the heating barrel, forming voids on the surface of the molded part. Groups of voided areas will gather to form a dull area because the underlying plastic cannot replicate the steel surface of the cavity.

Solution: Dry the material before processing. All materials require drying, even if they are not hygroscopic. After drying, the material must be used within two hours or moisture may return.

OPERATOR

Inconsistent Process Cycle

Explanation: It is possible that the machine operator is the cause of delayed or inconsistent cycles. This will result in erratic heating of the material in the injection barrel. If such a condition exists, some of the material may not absorb enough heat to travel far enough to pack the cavity and low gloss will occur because the material cannot replicate the steel surface.

Solution: If possible, operate the machine on automatic cycle, using the operator only to interrupt the cycle if an emergency occurs. Use a robot if an "operator" is really necessary. And, instruct all employees on the importance of maintaining consistent cycles.

Chapter 25
JETTING

Jetting can be defined as a "snake-like" pattern on the surface of a molded part, usually emanating from the gate area. It is indicative of an undesirable turbulence and splitting of the flow front.

Figure 25-1 *Jetting.*

The following can cause jetting:
- Excessive injection speed,
- barrel temperature too high or too low,
- small nozzle opening,
- low nozzle temperature,
- low mold temperature,
- small gates and/or runners,
- improper gate location,
- excessive gate land length,
- improper flow rate,
- inconsistent process cycle.

MACHINE

Excessive Injection Speed

Explanation: Excessive injection speed (fill rate) will cause the molten plastic to form jet streams as it is pushed through the gates instead of the more desirable wide "tongue" of material. These snake-like streams cool independently from the surrounding material and are visible on the molded part surface.

Solution: Reducing the injection speed will allow the plastic flow front to stay together and not form the individual streams that cause the jetting patterns on the part surface.

Barrel Temperature Too High or Too Low

Explanation: When barrel temperatures are too low, the material will not be heated to the proper temperature for adequate flow. The material will push slowly into the mold and the flow front will break up into individual streams as resistance builds up. This will cause jetting patterns. On the other hand, if the barrel temperature is too high, the material is pushed into the mold too quickly. This causes the flow front to split apart as it enters the cavity and the jetting patterns will develop.

Solution: Decrease or increase the barrel temperature accordingly. The material supplier can recommend the best starting point for barrel temperature and it can be adjusted from that point. Make changes in 10 degree F increments and keep the profile so it is heating progressively from back to front.

Small Nozzle Opening

Explanation: If the nozzle opening (or sprue bushing opening) is too small for the material being molded, the restriction may cause the material to flow too slowly and solidify early. The flow front may break apart as it travels through the gate (due to sidewall friction) and jetting patterns may develop.

Solution: Increase the nozzle opening. As a general rule, the nozzle opening should never be less than 7/32" in diameter. The stiffer the material flow, the larger that opening should be. Make sure that the sprue bushing opening diameter matches or is 1/32" larger than the nozzle opening.

Low Nozzle Temperature

Explanation: As material is transported through the heating barrel, it is gradually brought up to the ideal processing temperature by absorbing heat from the heating bands and frictional heat, which is created by the shearing action of the rotating screw within the barrel. In the last heating zone, the material is exposed to is the nozzle. By the time the material gets to the nozzle, it should already be at ideal molding temperature and only a small amount of heat needs to be applied at this point to keep the resin flowing. If the nozzle is not hot enough, however, the material will begin to cool off too quickly as it leaves the barrel and the flow front will not flow properly, breaking into many streams and causing jetting.

Solution: Increase the nozzle temperature. As a rule-of-thumb the nozzle temperature should be set at 10 degrees F higher than the setting for the front zone of the barrel. This helps compensate for heat loss due to metal-to-metal contact between the nozzle and the sprue bushing and keeps the material hot enough to flow properly, eliminating jetting.

MOLD

Low Mold Temperature

Explanation: Generally, a hot mold will allow a material to stay molten longer than a cold mold and cause the molecules to pack together properly before they solidify. This results in a dense part with no separation of layers. If the mold is too cold, the molecules solidify before they are packed together and may break up into separate units. As they travel through the gate these units split up and form jetting patterns on the part surface.

Solution: Increase the mold temperature to the point at which the material has the proper flow and packs out the mold without jetting. Start with the material suppliers recommendations and adjust accordingly. Allow 10 cycles for every 10-degree change for the process to re-stabilize.

Small Gates and/or Runners

Explanation: Gates and/or runners that are too small will cause excessive restriction to the flow of the molten plastic. Many plastics will then begin to solidify before they fill the cavity. The result is an unpacked condition and the flow front may break into separate streams, causing jetting patterns to develop.

Solution: Examine the gates and runners to determine if any burrs or other obstructions exist. If possible, perform a computer analysis to determine the proper sizing and location of gates and runners. Ask the material supplier for data concerning gate and runner dimensioning for a specific material and flow rate.

Improper Gate Location

Explanation: If certain materials are injected directly across a flat cavity surface they tend to slow down quickly as a result of frictional drag and cool off before the cavity is properly filled. As the material tries to flow through the gate, it is pulled apart into several streams and this forms a jetting pattern on the part.

Solution: Relocate, or redesign, the gate so that the molten plastic is directed against an obstruction such as a core pin. This will cause the material to disperse and continue to flow instead of slowing down.

Excessive Gate Land Length

Explanation: The area that surrounds the gate itself is called its *land*. It determines the distance a material must travel in a restricted state immediately before it enters the cavity. The length of this travel (land) should be no more than 1/8". The land acts like a tunnel when the mold is closed and if the tunnel is too ling the material begins to cool off before it can get to the cavity. This causes the material to split into streams that create the familiar jetting pattern on the part.

Solution: Decrease the gate land length. It is best to construct the mold so that the gates are located in replaceable inserts. That way they can be replaced easily at times when adjustments are needed. The insert should include the land area. This land length should be no less than 0.030" and no greater than 0.125".

MATERIAL

Improper Flow Rate

Explanation: Resin manufacturers supply specific formulations in a range of standard flow rates. Thin-walled products may require an easy flow material while thick-walled products can use a material that is stiffer. It is better to use as stiff a flow as possible because that improves physical properties of the molded part. But, the stiff material will be more difficult to push and this may result in a breakup of the flow front as the material enters the gate. The breakup appears as a jetting pattern.

Solution: Utilize a material that has the stiffest flow possible without causing jetting. Contact the material supplier for help in deciding which flow rate should be used for a specific application.

OPERATOR

Inconsistent Process Cycle

Explanation: It is possible that the machine operator is the cause of delayed or inconsistent cycles. This will result in erratic heating of the material in the injection barrel. If such a condition exists, the colder particles may not fill the mold before they fully solidify. Jetting may be caused as these colder areas attempt to push through the gate and are torn apart due to sidewall friction.

Solution: If possible, run the machine on the automatic cycle, using the operator only to interrupt the cycle if an emergency occurs. Use a robot if an "operator" is necessary. In addition, instruct all employees on the importance of maintaining consistent cycles.

Chapter 26
KNIT LINES (WELDS)

Knit lines can be defined as the inability of two or more flow fronts to "knit" together during the molding process. This normally occurs around holes or obstructions and causes locally weak areas in the molded part.

Figure 26-1 *Knit lines.*

The following can cause knit lines:

- Low barrel temperature,
- inadequate back pressure,
- injection pressure or speed too low,
- low mold temperature,
- small gates and/or runners,
- improper gate location,
- excessive gate land length,
- improper flow rate,
- inconsistent process cycle.

MACHINE

Low Barrel Temperature

Explanation: When barrel temperatures are too low, the material will not be heated to the proper temperature for adequate flow. The material will push slowly into the mold and the flow fronts that form around obstructions will not be hot enough to reunite after they travel around those obstructions.

Solution: Increasing the barrel temperatures will allow the flow fronts to stay hotter longer and knit better when they reunite. It is practically impossible to eliminate knit lines once they are formed but they can be minimized.

Inadequate Back Pressure

Explanation: The back pressure control is used to impart a resistance to the molten material being prepared in the barrel for the upcoming cycle. This resistance is used to help preheat the material and also control the density of the melt before it is injected into the mold. If the back pressure setting is too low, the material may not be brought to the proper temperature and the knit line areas will be more evident due to their inability to reunite.

Solution: Increase the back pressure to raise the melt temperature and improve the ability for the fronts to unite. This is best accomplished by starting at the minimum of 50 psi and increasing in 10-psi increments until the knit lines improve. Do not exceed 300 psi. The higher the back pressure, the hotter the plastic, and excessive back pressure will thermally degrade the plastic.

Injection Pressure or Speed Too Low

Explanation: If the injection speed or pressure is too low, the molten plastic will not be pushed into the mold cavity fast enough to fill and knit at converging flow fronts because it cools down and solidifies too soon.

Solution: Increase the injection pressure or speed. While these two parameters are related, it is not proper to adjust them both at the same time. Adjust them each independently and monitor the results closely to determine whether or not the other needs adjustment. As a rule-of-thumb, it is best to make adjustments in increments of no more than 10% of the original setting.

MOLD

Low Mold Temperature

Explanation: Generally, a hot mold will allow a material to stay molten longer than a cold mold and cause the molecules to flow farther before they solidify. This results in a dense part with maximum welding at knit line locations. If the mold is too cold, the molecules solidify before they are packed and the knit lines will be more evident and much weaker.

Solution: Increase the mold temperature to the point that the material has the proper flow and packs out the mold with maximum knit line strength. Start with the material suppliers recommendations and adjust accordingly. Allow 10 cycles for every 10-degree change for the process to re-stabilize.

Small Gates and/or Runners

Explanation: Gates and/or runners that are too small will cause excessive restriction to the flow of the molten plastic. Many plastics will then begin to solidify before they fill the cavity. The result is an unpacked condition and knit lines will be weaker and more evident.

Solution: Examine the gates and runners to determine if any burrs or other obstructions exist. If possible, perform a computer analysis to determine the proper sizing and location of gates and runners. Ask the material supplier for data concerning gate and runner dimensioning for a specific material and flow rate.

Improper Gate Location

Explanation: If certain materials are injected directly across a flat cavity surface they tend to slow down quickly because of frictional drag and cool off before the cavity is properly filled. The flow front breaks into many streams and they have difficulty welding back together before they solidify.

Solution: Relocate, or redesign, the gate so that the molten plastic is directed against an obstruction such as a core pin. This will cause the material to disperse and continue to flow instead of slowing down.

Excessive Gate Land Length

Explanation: The area that surrounds the gate itself is called its land. It determines the distance a material must travel in a restricted state immediately before it enters the cavity. The length of this travel (land) s hould be no more than 1/8". The land acts like a tunnel when the mold is closed and if the tunnel is too long the material begins to cool off before it can get to the cavity. This causes the material to split into streams that can create knit line conditions.

Solution: Decrease the gate land length. It is best to construct the mold so that the gates are located in replaceable inserts. That way they can be replaced easily at times when adjustments are needed. The insert should include the land area. This land length should be no less than 0.030" and no greater than 0.125".

MATERIAL

Improper Flow Rate

Explanation: Resin manufacturers supply specific formulations in a range of standard flow rates. Thin-walled products may require an easy flow material while thick-walled products can use a material that is stiffer. It is better to use as stiff a flow as possible because that improves physical properties of the molded part. But the stiff material will be more difficult to push and this may result in a breakup of the flow front as the material enters the gate. This may cause streams of flow that cannot weld back together properly and they will form weak knit lines.

Solution: Utilize a material that has the stiffest flow possible without causing knit lines. Contact the material supplier for help in deciding which flow rate should be used for a specific application.

OPERATOR

Inconsistent Process Cycle

Explanation: The machine operator may be the cause of delayed or inconsistent cycles. This will result in erratic heating of the material in the injection barrel. If such a condition exists, the colder particles may not fill the mold before they fully solidify. The flow front may break up into streams that cannot weld back together properly and weak knit lines may form.

Solution: If possible, run the machine on the automatic cycle, using the operator only to interrupt the cycle if an emergency occurs. Use a robot if an "operator" is necessary. In addition, instruct all employees on the importance of maintaining consistent cycles.

Chapter 27
NON-FILL (SHORT SHOTS)

Non-fill can be defined as an incomplete molded part caused by insufficient material injected into the mold. Non-fill can be an extension of a flow line or knit line condition.

Figure 27-1 *Non-fill.*

The following can cause non-fill:

- Insufficient material in hopper,
- low barrel temperature,
- inadequate back pressure,
- injection pressure or speed too low,
- excessive non-return valve clearance,
- bridging in feed throat,
- insufficient barrel capacity,
- low mold temperature,
- small gates and/or runners,
- improper gate location,
- insufficient venting,
- improper flow rate,
- inconsistent process cycle.

MACHINE

Insufficient Material in Hopper

Explanation: The most common cause of non-fill is insufficient material being injected into the mold, caused by not enough material in the hopper. Normally this is the result of employee error and the hopper was not checked often enough. It simply ran out of material.

Solution: It is best to use an automated system that replenishes material in the hopper as it is used. That way the machine will never run out of plastic. If manual systems are utilized instead, the employee in charge must understand the importance of keeping the hopper filled. Alarm units can be used to emit an audible signal when the material in the hopper reaches a preset level.

Low Barrel Temperature

Explanation: When barrel temperatures are too low, the material will not be heated to the proper temperature for adequate flow. The material will push slowly into the mold and the flow fronts that form will not be hot enough to complete the filling of the mold.

Solution: Increasing the barrel temperatures will allow the flow fronts to stay hotter longer and complete the filling of the mold. Make sure the proper profile is being used and that the material heats progressively as it travels through the barrel from rear to front.

Inadequate Back Pressure

Explanation: The back pressure control is used to impart a resistance to the molten material being prepared in the barrel for the upcoming cycle. This resistance is used to help preheat the material and also control the density of the melt before it is injected into the mold. If the back pressure setting is too low, the material may not be brought to the proper temperature and the flow fronts will not travel as far resulting in non-fill.

Solution: Increase the back pressure to increase the melt temperature and improve the ability for the fronts to flow. This is best accomplished by starting at the minimum of 50 psi and increasing in 10-psi increments until the proper flow is attained. Do not exceed 300 psi. The higher the back pressure the hotter the plastic, and excessive back pressure will thermally degrade the plastic.

Injection Pressure or Speed Too Low

Explanation: If the injection speed or pressure is too low, the molten plastic will not be pushed into the mold cavity fast enough to fill because it cools down and solidifies too soon.

Solution: Increase the injection pressure or speed. While these two parameters are related, it is not proper to adjust them both at the same time. Adjust them each independently and monitor the results closely to determine whether or not the other needs adjustment. As a rule-of-thumb, it is best to make adjustments in increments of no more than 10% of the original setting.

Excessive Non-Return Valve Clearance

Explanation: The non-return valve, found in the front section of the screw and barrel assembly, keeps molten plastic from slipping back over the injection screw when the screw is pushed forward during the injection phase of the process. The valve lies between the outside diameter of the screw and the inside diameter of the barrel and creates a seal between the two. If there is too much clearance (due to wear), the sealing effect is lost and slippage occurs. This results in a reduction in volume of plastic that gets injected into the mold and non-fill occurs.

Solution: Inspect the non-return valve mechanism and replace worn or damaged components. This wear is normal but is accelerated by molding materials that have reinforcements (such as glass) in them. The valve should be inspected at least every three months.

Bridging In Feed Throat

Explanation: As material is fed from the hopper to the heating cylinder, it passes through the feed throat of the machine. This area must have a temperature maintained at around 100 to 150 degrees F, depending on the plastic being molded, to prepare the material for higher heats in the barrel. But, if the temperature is too high, the plastic pellets begin to get sticky and bond together. This will form a blockage in the feed throat and material will not be allowed to fall through. It forms a bridge across the feed throat opening.

Solution: Decrease the feed throat temperature. The material supplier can provide the proper temperature value for a specific material. Make sure there is no obstruction in the water line used for cooling the feed throat.

Insufficient Barrel Capacity

Explanation: If the mold is running in a machine that does not have a large enough barrel, the material in the barrel may not be allowed to stay there long enough to absorb enough heat. The cold material will not flow as far as a hot material would and non-fill will occur. Increasing the heat may only degrade the plastic.

Solution: Place the mold in a machine that utilizes the 20% to 80% rule. This states that, ideally, a barrel should be sized such that 50% of the capacity is used every shot, but based on heat sensitivity of the material being used, that ratio can be between 20% for most materials and 80% for heat sensitive materials. This formula allows enough time for the material to absorb heat properly before being molded.

MOLD

Low Mold Temperature

Explanation: Generally, a hot mold will allow a material to stay molten longer than a cold mold and cause the molecules to flow farther before they solidify. This results in a dense part with maximum fill. If the mold is too cold, the molecules solidify before they are packed and the flow fronts will not travel far enough to fill the cavity image.

Solution: Increase the mold temperature to the point at which the material has the proper flow and packs out the mold with maximum fill. Start with the material suppliers recommendations and adjust accordingly. Allow 10 cycles for every 10-degree change for the process to re-stabilize.

Small Gates and/or Runners

Explanation: Gates and/or runners that are too small will cause excessive restriction to the flow of the molten plastic. Many plastics will then begin to solidify before they fill the cavity.

Solution: Examine the gates and runners to determine if any burrs or other obstructions exist. If possible, perform a computer analysis to determine the proper sizing and location of gates and runners. Ask the material supplier for data concerning gate and runner dimensioning for a specific material and flow rate.

Improper Gate Location

Explanation: If certain materials are injected directly across a flat cavity surface, they tend to slow down quickly as a result of frictional drag and cool off before the cavity is properly filled. The flow fronts have difficulty traveling far enough to pack out the part.

Solution: Relocate, or redesign, the gate so that the molten plastic is directed against an obstruction such as a core pin. This will cause the material to disperse and continue to flow instead of slowing down.

Insufficient Venting

Explanation: Venting is used to remove trapped air and gases from the closed mold, so molten material will be able to flow into every section of the mold. If the air and gases are not removed they act as a barrier to the flow of the plastic and will not allow filling to occur.

Solution: Vent the mold by grinding thin (0.0005"-0.002") pathways on the shutoff area of the cavity blocks. The viscosity of the plastic being molded determines the depth of the vent. Stiff materials can utilize deeper vents but fluid materials require thinner vents. In either case, the concept is to remove air from the mold as fast as possible with as deep a gate as the material viscosity will allow.

MATERIAL

Improper Flow Rate

Explanation: Resin manufacturers supply specific formulations in a range of standard flow rates. Thin-walled products may require an easy flow material while thick-walled products can use a material that is stiffer. It is better to use as stiff a flow as possible because that improves physical properties of the molded part. But, the stiff material will be more difficult to push and this may result in the flow fronts not traveling far enough to fill the cavity image.

Solution: Utilize a material that has the stiffest flow possible without causing non-fill. Contact the material supplier for help in deciding which flow rate should be used for a specific application.

OPERATOR

Inconsistent Process Cycle

Explanation: It is possible that the machine operator is the cause of delayed or inconsistent cycles. This will result in erratic heating of the material in the injection barrel. If such a condition exists, the colder particles may not flow properly and the flow front may not be allowed to travel far enough to fill the mold.

Solution: If possible, run the machine on the automatic cycle, using the operator only to interrupt the cycle if an emergency occurs. Use a robot if an "operator" is necessary. In addition, instruct all employees on the importance of maintaining consistent cycles.

Chapter 28
SHRINKAGE (EXCESSIVE)

Excessive shrinkage can be defined as an extreme decrease in the dimensions of a molded part after it has cooled to room temperature.

Figure 28-1 *Shrinkage (excessive).*

The following can cause excessive shrinkage:
- Barrel temperature too high,
- insufficient injection pressure or time,
- inadequate cooling time ,
- insufficient cushion and/or hold time,
- excessive non-return valve clearance,
- mold temperature too high,
- small gates and/or runners,
- improper gate location,
- improper flow rate,
- inconsistent process.

MACHINE

Barrel Temperature Too High

Explanation: If the barrel temperature is too high, the resin absorbs an excessive amount of heat and this increases the size of the voided area between the plastic molecules. Upon cooling, the skin of the material solidifies first and the remaining resin closes up the excessively large molecules and voids as it cools, pulling the solidified skin with it. The larger the molecules and voids, the greater the amount of shrinkage.

Solution: Decreasing the barrel temperature allows the molecules and voids to expand and contract normally and provide consistent shrinkage values. Shrinkage is impossible to predict accurately, but keeping the material and mold temperatures at the right settings (as recommended by the material suppliers) will minimize the effective shrinkage.

Insufficient Injection Pressure or Time

Explanation: Injection pressure must be high enough to push molten material into the mold, through the runners and gates, and into the cavity image area. It should be used to force material into every part of the mold until it is packed solidly. The proper amount of pressure held for the proper amount of time ensures that all the resin molecules are held closely together while they cool and solidify. This minimizes the amount of shrinkage that will take place after the part is removed from the mold. But, if inadequate pressure is used or if it is applied for too short a period of time, the molecules will not be constrained during the solidification phase and the entire part will shrink excessively after removal from the mold.

Solution: Increase the amount of pressure or the time applied. Upon initial startup, the mold should be filled incrementally starting with intentional short shots (if the mold design allows) and progressively increasing pressure shot-by-shot until the mold is filled and packed properly. Then, parts should be inspected for critical dimensions. If the mold is new, and dimensions are incorrect, the mold should be returned to the moldmaker for adjustments. If the mold has already been in production and the dimensions are wrong, process parameters can be adjusted to make the part shrink less or more, whichever is required.

Inadequate Cooling Time

Explanation: The cooling phase of the total molding cycle determines how long the molten material is held in a constrained condition until a strong skin is formed. After that, the part can be ejected from the mold and the next cycle can begin. If that cooling time is too short, the skin will not be of sufficient thickness and strength to hold the part together after ejection from the mold and continued shrinking may become excessive.

Solution: Adjust the cooling time portion of the cycle. It is true that longer cycles make the part more expensive but there is a minimum amount of time required for the resin to form a proper skin. It depends on what material is being molded. A general rule-of-thumb for a part with a wall thickness of 0.100", the cooling time should be 20 seconds. The overall cycle would then be 25-30 seconds.

Insufficient Cushion and/or Hold Time

Explanation: A cushion (pad) is required at the end of the injection stroke to maintain steady pressure against the molten material after it has filled the cavity image. This pressure (and the time it is applied) keep the molecules packed together and constrained while the product skin forms and solidifies enough for the part to be ejected. If the cushion is too small or the time is too short, the plastic molecules are not constrained and some will actually pull back out of the cavity. The remaining molecules may move after the part is ejected and cause excessive shrinkage as they contract beyond their normal values.

Solution: Maintain a cushion that is somewhere between 1/8" and 1/4" at the end of the injection stroke. This provides something against which hold pressure can be applied. The amount of time for holding pressure should be long enough for the gate to freeze, normally 3-4 seconds.

Excessive Non-Return Valve Clearance

Explanation: The non-return valve, found in the front section of the screw and barrel assembly, keeps molten plastic from slipping back over the injection screw when the screw is pushed forward during the injection phase of the process. The valve lies between the outside diameter of the screw and the inside diameter of the barrel and creates a seal between the two. If there is too much clearance (due to wear) the sealing effect is lost and slippage occurs. This results in a reduction in volume of plastic and a bottoming out of the injection stroke, eliminating the required cushion.

Solution: Inspect the non-return valve mechanism and replace worn or damaged components. This wear is normal but is accelerated by molding materials that have reinforcements (such as glass) in them. The valve should be inspected at least every three months.

MOLD

Mold Temperature Too High

Explanation: Generally, a hot mold will allow a material to stay molten longer than a cold mold and cause the molecules to stay fluid longer before they cool and solidify. Upon ejection from the mold the material will be allowed to contract more than normal and excessive shrinkage will occur.

Solution: Increase the mold temperature to the point at which the material has the proper flow and packs out the mold with maximum fill. Start with the material suppliers recommendations and adjust accordingly. Allow 10 cycles for every 10-degree change for the process to re-stabilize.

Small Gates and/or Runners

Explanation: Gates and/or runners that are too small will cause excessive restriction to the flow of the molten plastic. Many plastics will then begin to solidify before they fill the cavity. The result is a material that is not fully contained within the metal mold surfaces and is allowed to shrink beyond normal values.

Solution: Examine the gates and runners to optimize their size and shape. Do not overlook the sprue bushing as a long sprue may solidify too soon. Use a heated bushing or extended nozzle to minimize sprue length. Ask the material supplier for data concerning gate and runner dimensioning for a specific material and flow rate.

Improper Gate Location

Explanation: If certain materials are injected directly across a flat cavity surface they tend to slow down quickly as a result of frictional drag and cool off before the cavity is properly filled. The material is not held under proper pressure while solidifying and excessive shrinkage will occur after ejection from the mold.

Solution: Relocate or redesign the gate so that the molten plastic is directed against an obstruction such as a core pin. This will cause the material to disperse and continue to flow instead of slowing down.

MATERIAL

Improper Flow Rate

Explanation: Resin manufacturers supply specific formulations in a range of standard flow rates. Thin-walled products may require an easy flow material while thick-walled products can use a material that is stiffer. It is better to use as stiff a flow as possible because that improves physical properties of the molded part. But the stiff material will be more difficult to push and this may result in a less dense material filling the cavity image. The lower this density, the higher the amount of shrinkage that will occur after ejection.

Solution: Utilize a material that has the stiffest flow possible without *causing* non-fill. Contact the material supplier for help in deciding which flow rate should be used for a specific application.

OPERATOR

Inconsistent Process Cycle

Explanation: The machine operator may be opening the gate too soon, thereby effectively shortening the overall cycle time. This would cause the part to be ejected before the skin has formed properly and excessive shrinkage may occur.

Solution: If possible, operate the machine on automatic cycle, using the operator only to interrupt the cycle if an emergency occurs. Use a robot if an "operator" is really necessary. And, instruct all employees on the importance of maintaining consistent cycles.

Chapter 29
SINK MARKS

A *sink mark* can be defined as a depression, resembling a dimple or groove, caused by excessive localized shrinking of the material after the part has cooled.

Figure 29-1 *Sink marks.*

The following can cause sink marks:
- Barrel temperature too high,
- insufficient injection pressure or time,
- inadequate cooling time,
- insufficient cushion and/or hold time,
- excessive non-return valve clearance,
- mold temperature too high opposite ribs,
- small gates and/or runners,
- improper gate location,
- excessive thickness at mating walls,
- improper flow rate,
- inconsistent process cycle.

MACHINE

Barrel Temperature Too High

Explanation: If the barrel temperature is too high, the resin absorbs an excessive amount of heat and this increases the size of the voided area between the plastic molecules. Upon cooling, the skin of the material solidifies first and the remaining resin closes up the excessively large molecules and voids as it cools, pulling the solidified skin with it. The larger the molecules and voids, the greater the amount of shrinkage. This is displayed as sink marks.

Solution: Decreasing the barrel temperature allows the molecules and voids to expand and contract normally and provide consistent shrinkage values. Consistent, uniform shrinkage minimizes the condition that causes sink marks.

Insufficient Injection Pressure or Time

Explanation: Injection pressure must be high enough to push molten material into the mold, through the runners and gates, and into the cavity image area. It should be used to force material into every part of the mold until it is packed solidly. The proper amount of pressure held for the proper amount of time ensures that all the resin molecules are the same size and are held closely together while they cool and solidify. This creates uniform shrinkage after the part is removed from the mold. But, if inadequate pressure is used or if it is applied for too short a period of time, the molecules will not be constrained during the solidification phase and the entire part will not shrink uniformly, resulting in sink marks.

Solution: Increase the amount of pressure or the time applied. Upon initial startup the mold should be filled incrementally starting with intentional short shots (if the mold design allows) and progressively increasing pressure shot-by-shot until the mold is filled and packed properly. If injection pressure and time are adequate the shrinkage should be uniform and consistent resulting in parts without sink marks.

Inadequate Cooling Time

Explanation: The cooling phase of the total molding cycle determines how long the molten material is held in a constrained condition until a strong skin is formed. After that, the part can be ejected from the mold and the next cycle can begin. If that cooling time is too short, the skin will not be of sufficient thickness and strength to hold the part together after ejection from the mold and continued cooling may cause sink marks due to inconsistent and non-uniform shrinkage.

Solution: Increase the cooling time portion of the cycle. It is true that longer cycles make the part more expensive but there is a minimum amount of time required for the resin to form a proper skin. It depends on what material is being molded. A general rule-of-thumb for a part with a wall thickness of 0.100", the cooling time should be 20 seconds. The overall cycle would then be 25-30 seconds.

Insufficient Cushion and/or Hold Time

Explanation: A cushion (pad) is required at the end of the injection stroke to maintain steady pressure against the molten material after it has filled the cavity image. This pressure (and the time it is applied) keep the molecules packed together and constrained while the product skin forms and solidifies enough for the part to be ejected. If the cushion is too small or the time is too short, the plastic molecules are not constrained and some will actually pull back out of the cavity. This allows the remaining molecules to move when the part is ejected and they will cause excessive shrinkage and sink marks as they contract beyond their normal values.

Solution: Maintain a cushion that is somewhere between 1/8" and 1/4" at the end of the injection stroke. This provides something against which hold pressure can be applied. The amount of time for holding pressure should be long enough for the gate to freeze, normally 3-4 seconds.

Excessive Non-Return Valve Clearance

Explanation: The non-return valve, found in the front section of the screw and barrel assembly, keeps molten plastic from slipping back over the injection screw when the screw is pushed forward during the injection phase of the process. The valve lies between the outside diameter of the screw and the inside diameter of the barrel and creates a seal between the two. If there is too much clearance (due to wear) the sealing effect is lost and slippage occurs. This results in a reduction in volume of plastic and a bottoming out of the injection stroke, eliminating the required cushion.

Solution: Inspect the non-return valve mechanism and replace worn or damaged components. This wear is normal but is accelerated by molding materials that have reinforcements (such as glass) in them. The valve should be inspected at least every three months.

MOLD

Mold Temperature Too High Opposite Ribs

Explanation: Generally, a hot mold will allow a material to stay molten longer than a cold mold and cause the molecules to stay fluid longer before they cool and solidify. Upon ejection from the mold the material will be allowed to contract more than normal and excessive shrinkage will occur. This condition often occurs in the area of ribs because of the extra plastic in those areas, which requires more extensive cooling to maintain consistent shrinkage. Inconsistent shrinkage will result in sink marks.

Solution: Decrease the mold temperature to the point at which the material has the proper flow and packs out the mold without shorting. Start with the material suppliers recommendations and adjust accordingly. Allow 10 cycles for every 10-degree change for the process to re-stabilize.

Small Gates and/or Runners

Explanation: Gates and/or runners that are too small will cause excessive restriction to the flow of the molten plastic. Many plastics will then begin to solidify before they fill the cavity. The result is a material that is not fully contained within the metal mold surfaces and is allowed to shrink beyond normal expectations. The extended shrinkage causes sink marks.

Solution: Examine the gates and runners to optimize their size and shape. Do not overlook the sprue bushing as a long sprue may solidify too soon. Use a heated bushing or extended nozzle to minimize sprue length. Ask the material supplier for data concerning gate and runner dimensioning for a specific material and flow rate.

Improper Gate Location

Explanation: If certain materials are injected directly across a flat cavity surface they tend to slow down quickly as a result of frictional drag and cool off before the cavity is properly filled. The material is not held under proper pressure while solidifying and excessive shrinkage will cause sink marks as the part cools after ejection from the mold.

Solution: Relocate or redesign the gate so that the molten plastic is directed against an obstruction such as a core pin. This will cause the material to disperse and continue to flow instead of slowing down.

Excessive Thickness at Mating Walls

Explanation: When a wall meets another wall, or when a boss is located on a wall, the area where they form a junction becomes a larger mass of plastic than the area surrounding it. The surrounding area cools and is already solidified while the larger mass continues to cool and shrink. Because the surrounding area is solid, non-uniform shrinkage occurs as the large mass area shrinks in on itself, causing sink marks to appear.

Solution: Although it is good design practice to maintain all walls at a uniform thickness, in areas where a junction is formed, one of the walls should be between 60% and 70% of the mating wall thickness. This will minimize the mass at the junction until the shrinkage is equal in all areas and sink marks will not develop.

MATERIAL

Improper Flow Rate

Explanation: Resin manufacturers supply specific formulations in a range of standard flow rates. Thin-walled products may require an easy flow material while thick-walled products can use a material that is stiffer. It is better to use as stiff a flow as possible because that improves physical properties of the molded part. But the stiff material will be more difficult to push and this may result in a less dense material filling the cavity image. The lower t his density, the higher the amount of shrinkage that will occur after ejection, and sink marks may occur due to an imbalance of shrinkage factors.

Solution: Utilize a material that has the stiffest flow possible without causing sink marks. Contact the material supplier for help in deciding which flow rate should be used for a specific application.

OPERATOR

Inconsistent Process Cycle

Explanation: The machine operator may be opening the gate too soon, thereby effectively shortening the overall cycle time. This would cause the part to be ejected before the skin has formed properly and the resultant excessive shrinkage may cause sink marks to form.

Solution: If possible, run the machine on the automatic cycle, using the operator only to interrupt the cycle if an emergency occurs. Use a robot if an "operator" is necessary. In addition, instruct all employees on the importance of maintaining consistent cycles.

Chapter 30
SPLAY

Splay can be defined as a "splash-like" appearance (or spray pattern) on the surface of a molded part. Splay is sometimes called "silver-streaking."

Figure 30-1 *Splay.*

The following can cause splay:
- Barrel temperature too high,
- excessive screw rotation speed,
- nozzle too hot, too small, or obstructed,
- excessive shot size,
- trapped volatiles,
- improper purging,
- small gates,
- gate and/or runner obstruction,
- cracked mold,
- excessive moisture,
- inconsistent process cycle.

MACHINE

Barrel Temperature Too High

Explanation: If the barrel temperature is too high, the resin absorbs an excessive amount of heat and will decompose and/or carbonize. The charred molecules that are created will not bond with the surrounding material and will float to the surface during injection. This results in a spray of charred particles on the surface of the molded part, which are usually fanned out in direction emanating from the gate location.

Solution: Decreasing the barrel temperature allows the plastic to stay molten without burning or charring. The molecules bond together as desired and splay is eliminated.

Excessive Screw Rotation Speed

Explanation: Screw rotation is utilized to transfer material through the barrel from rear to front. Also, it is used to impart frictional heat to the resin, which helps to melt it. If the rotation is too fast, some of the material may overheat and degrade due to increased frictional shear. As it is injected into the mold, the degraded material pushed to the surface appears as splay.

Solution: Reduce the screw rotation speed. This will allow the material to be properly heated and blended without being degraded. A good starting point for screw rotation on a standard two-inch diameter screw is 120 rpm.

Nozzle Too Hot, Too Small, or Obstructed

Explanation: The nozzle, being the final transfer point between the heated barrel and the mold, is a critical area and must be scrutinized whenever splay patterns occur. If the nozzle is too hot, too small, or obstructed, the molten material may become degraded as it travels through the area and will surface in the mold cavity as splay. This is due to frictional shear heat generated at the point of obstruction or overheating.

Solution: Reduce the nozzle temperature 10 degrees F at a time until the splay disappears or the nozzle freezes off whichever comes first. If splay still appears, check for obstructions such as tramp metal or burrs. If splay is still evident, make sure the nozzle opening is large enough for the material being molded. The material supplier can provide this information. And, finally, make sure the nozzle being used is of the proper design for the material being molded. There are many different nozzle designs and some may interfere with proper flow if they are not designed for the material in use.

Excessive Shot Size

Explanation: Ideally, a shot size should be between 20% and 80% of the barrel capacity, with 50% being just right. The lower figure can be used for non-heat-sensitive materials, while the higher number should be used for very heat-sensitive materials. Using such a ratio will ensure that the residence time is not too great. If it is, the material will thermally degrade in the barrel and may appear as splay in the molded part.

Solution: Run the mold in a press that is sized to provide the proper ratio of shot-to-barrel size. This will minimize the risk of thermally degrading the material.

Trapped Volatiles

Explanation: During the plasticizing process, there are volatiles that are released from the melting plastic. If not properly removed, these gases will join to form pockets that are pushed into the melt stream and carried into the cavity where they show up on the molded part as splay.

Solution: Increase the back pressure setting to remove the volatiles. Higher back pressure settings help to disperse the volatiles throughout the melt stream and keep them from joining up to form pockets of trapped gas.

Improper Purging

Explanation: When switching materials on a specific press, there is a tendency to start molding with the new material without properly purging all of the older material out of the barrel. The older material may not mix with the newer material due to temperature differences or chemical incompatibility and a smearing effect may appear on the surface of the molded parts. This smearing is also considered to be splay.

Solution: Purge out the offending material using a minimum of 20 shots. Use a purging compound especially designed to clean barrels, or use scrapped acrylic, which tends to scrub the barrel clean.

MOLD

Small Gates

Explanation: Gates that are too small will cause excessive restriction to the flow of the molten plastic as it passes through. This restriction may cause enough shear heat to thermally degrade the plastic as it fills the cavity. The degraded material gets sprayed across the surface of the molded part as splay.

Solution: Optimize gate size and shape. The material supplier can provide data on proper sizing and shape, or use a computer finite element analysis program to help make the determination.

Gate and/or Runner Obstruction

Explanation: On new molds, a machining burr may have been left on the parting line. Or, on older molds the parting line surface may have started to peen over. In either case, an obstruction forms that will interfere with the material flow going through the runners and/or gates and cause an overheating condition due to excessive shearing action. This causes some of the material to be thermally degraded and it is molded into the part as splay.

Solution: Check the parting line and all other areas surrounding the runners and gates. If obstructions exist remove them. Peened edges and burrs can be stoned and polished while cracks and nicks may have to be welded and recut or inserts can be made to replace those areas.

Cracked Mold

Explanation: If the mold base is cracked it is possible that water from cooling lines can enter the cavity. This water gets molded into the part and turns to steam as the molten plastic captures it. The steam will be sprayed out across the cavity surface as splay.

Solution: Inspect for cracks, even small cracks, and repair as necessary. If the cracks are in areas that are too critical to weld, it is possible to thread tubing through the cooling channels and let the water travel through the tubing. This will prevent the water from entering the cavity area.

MATERIAL

Excessive Moisture

Explanation: Excessive moisture does not belong in the molding compound. Moisture turns to steam when heated in the injection unit, and these steam pockets interfere with molecular bonding of the plastic. This causes splay, which is a visual defect, but also creates a weak part due to brittleness.

Solution: Although it is commonly understood that non-hygroscopic material does not require drying, do not take chances. Dry all materials. It may be that fillers used in the material **are** hygroscopic and they will absorb moisture. Every plastic material requires specific drying conditions. And each material should be dried according to the material suppliers recommendations. The desired moisture content is between 1/10th of 1 percent and 1/20th of 1 percent by weight. This means the dry air being used to take moisture from the material should have a dew point of -20 to -40 degrees F.

OPERATOR

Inconsistent Process Cycle

Explanation: The machine operator may be the cause of delayed or inconsistent cycles. This will result in erratic heating of the material in the injection barrel. If such a condition exists, some of the material will overheat and degrade. It may end up being carried into the cavity and sprayed across the molded part surface as splay.

Solution: If possible, run the machine on the automatic cycle, using the operator only to interrupt the cycle if an emergency occurs. Use a robot if an "operator" is necessary. In addition, instruct all employees on the importance of maintaining consistent cycles.

Chapter 31
WARPAGE

Warpage can be defined as a dimensional distortion in a molded product after it is ejected from the mold at the end of the injection molding process. Warpage is sometimes called "potato-chipping" because the part tends to appear wavy.

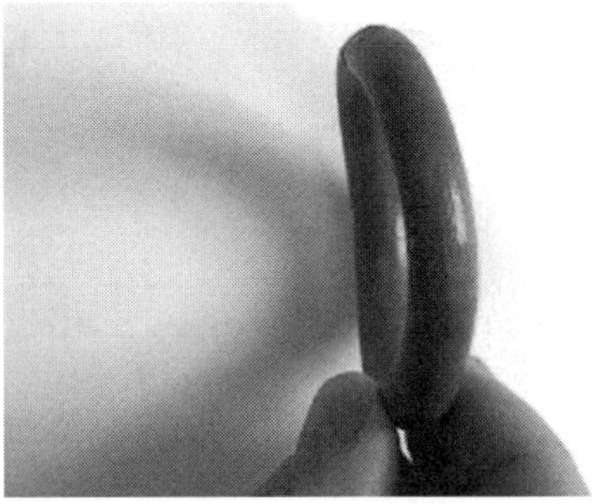

Figure 31-1 *Warpage.*

The following can cause warpage:
- Inadequate injection pressure or time,
- inadequate residence time,
- barrel temperature too low,
- nozzle temperature too low,
- excessive stress buildup,
- small gates,
- mold temperature too low,
- uneven mold temperatures,
- non-uniform ejection,
- improper flow rate,
- inconsistent process cycle.

MACHINE

Inadequate Injection Pressure or Time

Explanation: If too little injection pressure is used the plastic material will tend to cool and solidify before the mold is packed out. If no packing is achieved the individual molecules are not held tightly together and have space to move while the part is cooling. Also, if the injection hold time is not long enough, the packing process is minimized and the molecules can relax before full solidification occurs. In either case, as the part cools it is uncontrolled and the plastic is allowed to move because it is not being constrained. Each area of the part cools at a different rate and warpage will occur due to the differences.

Solution: Increase the injection pressure or time applied. This will ensure the total part is cooling while constrained and the tendency for warpage will be minimized.

Inadequate Residence Time

Explanation: Residence time is the amount of time a material must spend being exposed to heat in the barrel. The time is determined by the ability of the specific resin to absorb heat enough to be properly processed. Inadequate residence time will result in under-heated material, which causes the material to be stiff. It will cool off before the mold is packed and individual molecules will be unconstrained while they solidify. Molecules that are not constrained during cooling will shrink at differing rates throughout the part and warpage will occur.

Solution: Increase the residence time by adding time to the cooling portion of the cycle. While increased cycle time may add cost to the final product, each material requires a specific minimum amount of time to absorb heat in the barrel, and if the time is not long enough warped parts will occur.

Barrel Temperature Too Low

Explanation: When barrel temperatures are too low, the material will not have a chance to heat to the proper flow temperature. The cold material gets pushed into the mold but solidifies before the molecules are packed and constrained. This results in warpage as the molecules shrink at varying rates.

Solution: Increase the barrel temperature. This will allow the material to come to proper heat and it will fill the mold before solidification takes place. The molecules will be packed and constrained as they cool, thus shrinking at uniform rates, minimizing the chance of warpage.

Nozzle Temperature Too Low

Explanation: The nozzle, being the final transfer point between the heated barrel and the mold, is a critical area and must be scrutinized whenever splay patterns occur. If the nozzle is too cold, the plastic material may slow down as it travels through the area and the molecules will not get packed under constraint. They will shrink at varying rates and cause warpage.

Solution: Increase the nozzle temperature 10 degrees F at a time until the warpage disappears. If splay still appears, reduce the temperature of the nozzle, and make sure the nozzle being used is of the proper design for the material being molded. There are many different nozzle designs and some may interfere with proper flow if they are not designed for the material in use.

Excessive Stress Buildup

Explanation: The injection molding process tends to build up physical stress in a molded part due to the stretching and squeezing action that takes place on the individual plastic molecules as they are heated, expanded, cooled and contracted. They must be allowed to relax and recover in a constrained position before they solidify or the stress will be locked in. It will then be released as the part cools after being ejected from the mold and warpage will occur.

Solution: Increase the barrel temperature and decrease injection pressure until the stress is minimized. It can never be eliminated but lower pressure will result in lower stress. And, higher barrel temperatures allow the use of lower injection pressures.

MOLD

Small Gates

Explanation: Gates that are too small will cause excessive restriction to the flow of the molten plastic as it passes through. This restriction may cause additional physical stress to the plastic molecules as they are stretched and squeezed again going through the gate area. The stress gets released after ejection and the parts will warp.

Solution: Optimize gate size and shape. The material supplier can provide data on proper sizing and shape, or use a computer finite element analysis program to help make the determination.

Mold Temperature Too Low

Explanation: Generally, a hot mold will allow a material to stay molten longer than a cold mold and cause the molecules to stay fluid longer before they cool and solidify. If the mold is too cold the molecules will solidify before they are packed and will shrink at differing, uncontrolled rates. This is a prime cause of warpage.

Solution: Increase the mold temperature to the point at which the material has the proper flow and packs out the mold with maximum fill. Start with the material suppliers recommendations and adjust accordingly. Allow 10 cycles for every 10-degree change for the process to re-stabilize.

Uneven Mold Temperatures

Explanation: The plastic molecules must cool and shrink evenly to resist warpage. If the mold is not cooling the plastic in a uniform manner the molecules will have varying cooling and shrinking characteristics and this will cause warpage.

Solution: Check the surfaces of the mold that are in contact with the molten plastic. Use a fast-acting pyrometer to determine if there is more than a 10 degree F difference between any two points, even between the two mold halves. A difference greater than 10 degrees F will cause too great a difference in shrink rates and warpage will occur.

Non-Uniform Ejection

Explanation: It is possible that either the ejection system of the mold or the press will not be operating properly. If the part is warm enough and the ejection force is not even and exactly perpendicular to the part, stresses will be set up in the part as it tries to resist the ejection activity. These stresses will cause warpage of the part as it cools after being ejected.

Solution: Inspect and adjust the ejection system(s) as required. Make sure all adjusting devices are locked down to eliminate slipping, and that all components are properly lubricated. It may be necessary to use a guided ejection system that utilizes leader pins and bushings to keep the system in line and even.

MATERIAL

Improper Flow Rate

Explanation: Resin manufacturers supply specific formulations in a range of standard flow rates. Thin-walled products may require an easy flow material while thick-walled products can use a material that is stiffer. It is better to use as stiff a flow as possible because that improves physical properties of the molded part. But the stiff material will be more difficult to push and this may result in the material solidifying before full packing takes place. The molecules will be left to shrink at different rates and warpage will occur.

Solution: Utilize a material that has the stiffest flow possible without causing warpage. Contact the material supplier for help in deciding which flow rate should be used for a specific application.

OPERATOR

Inconsistent Process Cycle

Explanation: The machine operator may be opening the gate too soon, thereby effectively shortening the overall cycle time. This would cause the part to be ejected before the skin has formed properly and excessive, uncontrolled shrinkage may occur. The varying shrinkage rates will cause warpage.

Solution: If possible, run the machine on the automatic cycle, using the operator only to interrupt the cycle if an emergency occurs. Use a robot if an "operator" is necessary. In addition, instruct all employees on the importance of maintaining consistent cycles.

INDEX

D

E

F

G

H

I

J

K

L

V

W